Ulrich Hauptmanns · Wolfgang Werner

Engineering Risks

Evaluation and Valuation

With 52 Figures

Springer-Verlag Berlin Heidelberg NewYork
London Paris Tokyo Hong Kong Barcelona

Dr. Ing Ulrich Hauptmanns
Dr. rer. nat. Wolfgang Werner

Gesellschaft für Reaktorsicherheit
GRSmbH
Schwertnergasse 1
5000 Köln 1

This is a completely revised edition on the basis of
Technische Risiken,
1st edition, Springer-Verlag 1987

ISBN-13: 978-3-642-95612-6   e-ISBN-13: 978-3-642-95610-2
DOI: 10.1007/978-3-642-95610-2

Library of Congress Cataloging-in-Publication Data
Hauptmanns, Ulrich.
[Technische Risiken. English]
Engineering risks--evaluation and valuation / U. Hauptmanns, W. Werner.
Translation of: Technische Risiken.
Includes bibliographical references and index.

1. Technology--Risk assessment. 2. Reliability (Engineering)
I. Werner, Wolfgang. II. Title.
T174.5.H3813   1991
620'.004--dc20   90-25317

The use of registered names, trademarks, etc. in this publication does not imply, even in the absence of a specific statement, that such names are exempt from the relevant protective laws and regulations and therefore free for general use.

Offsetprinting: Mercedes-Druck, Berlin; Bookbinding: Lüderitz & Bauer, Berlin
61/3020-543210 – Printed on acid-free paper

Preface

Encouragement by colleagues and a considerable increase in the use of probabilistic analyses since the publication of the German edition in 1987 have motivated this English version.

A mere translation was inappropriate because a number of important studies completed in recent years had to be included, among them the assessment of the risks of five nuclear power plants in the United States of America and the German Risk Study, Phase B. The opportunity was taken to elaborate on some concepts which have gained importance of late such as accident management. An update of international safety goals was also made; however, this can only be a momentary view of a field subjected to frequent change.

Thanks are due to the Springer-Verlag for the careful editing and production of the book.

Köln, Garching
March 1990

Ulrich Hauptmanns
Wolfgang Werner

Preface to the German Edition

With the increasing use of complex technologies there is a growing need to evaluate the associated risks. The methodology of probabilistic safety and risk analysis allows predictive valuation of risks. Nuclear engineering has been in the forefront of the development and application of this method. In the Safety Study on US Power Plants published in 1975 the risk of an entire technology was investigated systematically and quantified for the first time. Meanwhile the methods have continuously been improved and applied to a number of nuclear power stations. Risk assessments have also been performed in other sectors of industry; for example, for process plants. It can be expected that risk studies will be employed more and more to support decisions on the use and further development of technologies with large hazard potentials.

This book describes the procedure for carrying out risk studies focussing on investigations for nuclear installations. Additionally, studies on process plants and risk comparisons for different technologies of energy generation are addressed. It becomes clear that results from many different fields of knowledge are required for carrying out a risk study. The decisive advantage of such an investigation is that these results are combined into a global result. But it also becomes clear that despite intense research there remain phenomena which are not fully understood and that uncertainties always affect the results of risk studies.

The subject is described verbally so that reading the book requires no specific mathematical background. The results of numerous studies are presented. The understanding of the underlying concepts and procedures allows the reader to judge their value and usefulness.

The book is based on the state-of-the-art report on risk elaborated by the Gesellschaft für Reaktorsicherheit (GRS) in Cologne at the request of the Federal Minister of the Interior in August 1984. This report was revised and updated with exception of the part on probabilistic safety goals in foreign countries. In many cases these are not legally binding and therefore subject to frequent change.

In writing the book the authors received advice from a number of colleagues. They wish to express their gratitude to Messrs. Hörtner (dependent failures), Köberlein (studies on fast breeder and high temperature reactors), Polke (risk of the nuclear fuel cycle), and Reichart (human error). Thanks are also due to the Federal Ministry for the Environment, Nature Conservation, and Nuclear

Safety who made the publication possible, and to the Springer-Verlag for the good presentation of the book.

Bonn, Garching, Köln   The authors

Contents

1 Introduction

Risk, that is the possibility of suffering damage, is inseparably associated with human existence. On one hand it results from natural causes like illness or from natural disasters like earthquakes, floods or volcanic eruptions; on the other hand it is a side effect of man's technological achievements. The risk resulting from the utilization of technical installations is not only the subject of research in engineering itself, but also in jurisprudence and sociology [1 – 5]. In engineering the chief aim is to identify sources of risk, to quantify the risk resulting from them, and to develop and implement measures to reduce it. Legislation has the responsibility to protect man and property from the risks of technical installations; sociology deals with the effects of technology on human life forms and with the problems of acceptance of engineering risks.

This book is exclusively concerned with the technological aspects of risk; it deals mainly with nuclear power technology. Additionally, some studies on process plants are presented. Moreover, risk comparisons in the fields of nuclear and non-nuclear generation of electrical energy are considered. This limitation of scope does not imply the absence of risk considerations and of the use of probabilistic methods in other fields, like, for example, civil and aeronautical engineering, space technology [6], the transport and storage of dangerous substances [7], off-shore platforms [8], and biological [9] and medical sciences [10].

Before discussing the methods of risk assessment and the results of risk studies in detail, the fundamental notion of risk is clarified in Sect. 1.1, and the potential of probabilistic methods in safety and risk valuation is described in Sect. 1.2.

1.1 Definition of Risk

In everyday language the notion of risk is related to venture, hazard or danger, that is, the possibility to incur damage. In this sense risk is defined here as a quantitative measure of hazard.

Risk is also assessed in everyday life. For example, the risk of being hit and killed by a meteorite is considered to be extremely low, whereas the risk of

falling ill from influenza during the winter is considered to be high because the
event is very frequent.

Insurance companies use a more precise concept of risk. They estimate, for
example, the risk of damage due to fire or to burglary. On this basis they
calculate the premium to be paid by somebody who wants to be insured against
such risks.

Before assessing a risk it must be clearly defined and described which of the
many possible risks are of interest and are to be determined. In this context,
the following questions are asked:

– Whose risk is to be assessed?

For every individual or social group (family, community, economic enterprise,
nation, etc.) there exist various specific risks. When referring to a person we
speak of individual risk, in the case of several people the terms collective, societal, group or public risk are used. With an economic enterprise the investment
risk is considered.

– Which risk to a person or to the public is to be determined?

The risks may be of a very specific nature; for example, death due to lightning
or risks occurring in a more narrowly defined context like "working in the
household", "car traffic", or "operation of power plants".

– For what period of time is the risk to be determined?

A person or group is exposed to a risk as long as the possibility exists of the
damage in question occurring. Some possibilities exist permanently, others only
at certain times of the day or seasons of the year or only during clearly defined
activities, for example, during the take-off or landing of an aircraft.

Every possibility to suffer damage is a risk, if it is uncertain, whether it will
become reality. Damage and uncertainty therefore are the two elements which
determine a risk. If the magnitude of damage and the uncertainty of its occurrence may be quantified, then a number may be assigned to the corresponding
risk, the risk number. Frequently, this number refers to a certain period of
exposure, for example, one year. It is then called "risk per year".

1.1.1   The Element of "Damage"

In order to obtain a numerical value for the damage, it must be measurable,
that is to say, expressible in units of measure. In case of an accident causing
fatalities, damage is measured by the number of persons killed. Damage to
property, for instance due to a fire in a factory, is measured in monetary units.
For some categories of damage it is difficult to define measuring units; for
example, for damage to the environment or for psychic harm. Damages which
are described in terms of different measuring units belong to different damage

categories. In general, they are not comparable. Table 1.1 shows a few examples of different damage categories and the corresponding measuring units.

The extent of damage cannot always be determined exactly. This is true for events in the past whose precise description may imply problems of delineation, and in particular for the analytical modelling of events potentially causing damage. For example, it is not always possible to determine whether observed detrimental health effects had already existed previously or whether they were caused by the event under consideration. If a study requires the consequences of an explosion to be treated analytically, a model must be developed which is capable of describing the propagation of the pressure wave on the basis of chemical and physical laws and of accounting for such influential factors as local conditions and possibly existing obstacles. In general, with such a model the magnitude of damage can only be assessed approximately.

Table 1.1. Examples of measuring units of damages

Type of damage	Measuring unit
Human death	Number of deaths
Health effects	Number of affected persons (e.g. injured persons)
Regions rendered uninhabitable	Surface area
Material damage (replaceable)	Monetary units

1.1.2   The Element of "Uncertainty"

Uncertainty is expressed by the question "What is the probability of occurrence of an event?" Thus, we speak of probability as the uncertainty expressed in numbers. In mathematics the concept of probability has been precisely defined.

An impossible event has the probability 0 and an event with absolute certainty the probability 1. A possible, yet uncertain event has a probability between 0 and 1. The more probable the event, the closer is this number to 1.

Questions about probability are often difficult to answer, as is shown below:

Let X be a set of n elements (population) and $Y \subset X$ be a subset of X consisting of m elements with the characteristic y. The probability to obtain an element Y, if an element of X is taken randomly, is m/n. However, in practically important cases

– the number n of elements of X and
– the number m of elements with the characteristic y are unknown.

In order to determine approximately the unknown quotient m/n a sample, that is v elements of X are taken; then the number of elements u with the characteristic y is determined. The quotient u/v, the so-called relative frequency, is then used as an estimate for the unknown probability m/n.

This estimate is uncertain and the uncertainty can be quantified by determining an interval (the so-called confidence interval), which contains the correct probability with a certain level of confidence.

If probabilities derived from observations are to be used for predictions, then uncertainties frequently arise because the conditions under which the observations were made do not exactly apply to the period of time for which the predictions are to be made.

Probability values in practical risk calculations are, in general, only estimates. So-called objective estimates are based on samples from the exact population, for which the probability estimate is to be made. However, if the estimate is based on samples from different populations or on information whose suitability may be questioned, then the estimates are called subjective. They are then based on the personal judgement that the possible differences between the population from which the sample is taken and the population for which the estimate is to be made do not significantly affect the estimated value. This kind of subjective probability estimate is, particularly, necessary for predictions.

Consequently, we then speak of subjective confidence intervals and subjective confidence levels. Subjective estimates can provide reasonable values if the personal judgement involved is based on professional experience, that is, if it is a so-called "expert" judgement.

1.1.3   Combining Magnitude of Damage and Probability to Risk Numbers

1.1.3.1   Individual Risk

Let $y_1, y_2, \ldots$ be elements of a denumerable set of different magnitudes of damage per occurrence of an event belonging to a certain damage category, for example number of casualties, and $WS(y_i)$ the probabilities of their occurrence in the time interval under consideration. Then the individual risk caused by them is described by the risk number:

$$R^* = y_1 WS(y_1) + y_2 WS(y_2) + \ldots \tag{1.1}$$

The number R^* is an estimate of the risk if either the magnitudes of damage or their probabilities of occurrence or both are estimates. It coincides with the estimate of the mean individual probability of the event, if this can occur only once and if each occurrence can only cause a damage of magnitude 1.

Often the expected frequencies of occurrence $h(x_1), h(x_2), \ldots$ of events causing damage of magnitude $x_1, x_2, \ldots$ are used. Then, the risk number

$$R = x_1 h(x_1) + x_2 h(x_2) + \ldots \tag{1.2}$$

indicates the expected damage per year. If the magnitudes of damage per event are independent of each other, identically distributed, and independent of the number of occurrences of the event, then the risk number can also be determined according to the relationship $R = \bar{h}\bar{x}$. Here, $\bar{x}$ is the expected average magnitude of damage per event, and $\bar{h}$ is the mean value of its expected annual frequency of occurrence.

If the risk resulting from all events in a specified context, like for example the operation of a nuclear power plant, is to be assessed, then the risk numbers are added if the events are mutually exclusive.

1.1.3.2 Collective Risks for a Specific Category of Damage

In principle, the collective risk equals the product of the number of exposed individuals in the group and of the individual risk. Vice versa, an estimate of the individual risk may be obtained if the estimate of the collective risk is divided by the number of exposed individuals in the group.

Collective risks may be quantified using the arguments which led to the formulae (1.1) and (1.2) for the individual risk. Thus, for example, if the frequencies of occurrence of traffic accidents causing $x_1, x_2, \ldots$ fatalities in a country in the coming year were known, then the collective risk could be estimated according to relation (1.2).

Naturally we would not normally calculate the collective risk using this formula, because the number of traffic fatalities in any one year is already a useful estimate for the following year, within the achievable accuracy.

However, if we have to estimate risk numbers of events which have rarely or never been observed, so that figures from the past alone do not allow one to predict risk satisfactorily, the risk number can only be determined according to the formulae of types (1.1) and (1.2). This will be explained in more detail in the next section.

1.1.4 Risk Numbers for Rare or Unobserved Events

In Sects. 1.1.1 to 1.1.3 it was described how to combine the magnitude of damage (for a damage category) and its frequency to yield a risk number. Formally this is expressed by the simple relationships (1.1) and (1.2). By using these relations the risk numbers for different magnitudes of damage and all conceivable ways of damage occurring may be calculated. Naturally, all possible magnitudes of damage and their respective probabilities or frequencies of occurrence then have to be known. Unfortunately, this is normally not the case for rare events. In the following section we use simple examples to show how risk is quantified in such situations.

1.1.4.1   Risk Estimates Based on Observations of the Complementary Event

If the probability of occurrence of a rare event is to be estimated, it is an advantage if the so-called complementary event has frequently been observed. Commercial aircraft make several million take-offs and landings every year. The event "no crash during take-off or landing" is complementary to the event "crash during take-off or landing". If we assume, for example, that at airport X there have been 10 000 commercial take-offs or landings without a single crash, and if we take the first 1000 take-offs or landings as a random sample, then it does not contain the event "crash during take-off or landing". According to the classical methods of statistics we may then conclude that the probability of occurrence of the event "crash of a commercial aircraft at airport X" has an upper bound of 0.003 at a confidence level of 95%. Hence the corresponding confidence interval contains the values between 0 and 0.003. If not only the first 1000, but all 10 000 flight operations are used, then the upper 95% bound of the same event is 0.0003.

In the same way we could use operating experience with commercial light water reactors to estimate upper expected frequency bounds for certain events like a core melt. From 2200 reactor years (experience with light water moderated reactors with a power of 400 MWe and more in the Western world) with the occurrence of one core melt accident (partial core melt in TMI-2) we may conclude that the expected annual frequency of occurrence of that accident for each reactor of the above category lies below 0.0022 at a confidence level of 95%. This upper bound is determined as if we knew absolutely nothing about reactors except for the fact that in 2200 reactor years one such event has occurred. In contrast to the example of flight operations, no more observations are available to show that the upper bound is as high as it is only because of the relatively small number of observed reactor years. But even if a smaller upper bound could be estimated on the basis of additional observations of the complementary event, no statement could be made about the second element of the risk number, namely about the damage potentially caused by the event.

In such cases the events in question are decomposed into sub-events. This is done to a degree of detail such that

- partial events can be distinguished according to their influence on the magnitude of damage and
- sub-events of these partial events can be recognized, which can lead to the partial event only in combination with other sub-events and their probabilities can be reasonably estimated from observations and additional knowledge of details.

1.1.4.2   Risk Estimates Based on Knowledge of Details

Figure 1.1 outlines a risk assessment based on detailed knowledge proceeding from a description of the event to the determination of the damage and the risk numbers.

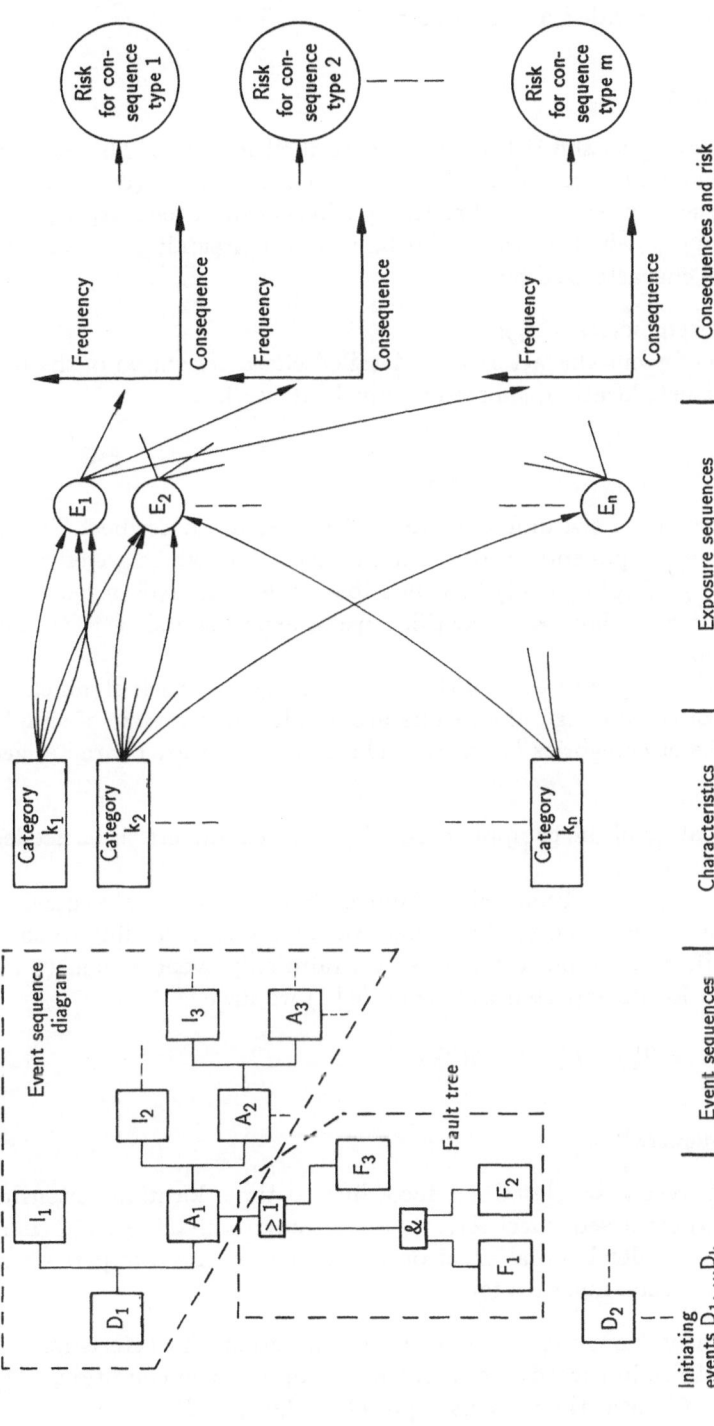

Fig. 1.1. Schematic diagram of a risk assessment based on detailed knowledge of event and exposure sequences (after [11])

The outline can be divided into four steps:

(a) "Event sequences"

All events contributing to the risk must be described in detail. This is done using event sequence diagrams in combination with fault trees (cf. Chap. 2). The various event sequences consist of concatenations of sub-events $A_1, A_2, \ldots,$ $I_1, I_2, \ldots,$ respectively which describe the failure or functioning of technical systems. They are characterized by

- their expected frequencies $h(T_i)$;
- the description of event characteristics (detailed characterization of the disturbance, i.e. cause, location, substances involved, etc.).

b) "Characteristics"

In this step the results of the different event sequences are described. This is done in terms of the components of the event characteristics which are essential to assess the damage (e.g. intensity of a possible explosion, level of heat and smoke generation etc.), which are quantified using experimental observations or model calculations.

Depending on the range of values of these components and their significance to an assessment of the damage, the results are divided for the sake of simplification into classes or categories $k_1, k_2, \ldots$. These categories are characterized by:

- representative values of the components of the event characteristic as needed to assess the damage, and
- the sums of the expected frequencies of mutually exclusive event sequences from (a), which are assigned to the respective categories according to their characteristic. If, for example, category k_1 contains only event sequences T_1, T_2, and T_5 then for its expected frequency $h(k_1)$ we have

$$h(k_1) = h(T_1) + h(T_2) + h(T_5) \tag{1.3}$$

(c) "Exposure sequences"

In this step all processes are described according to time, location, intensity and probability (exposure sequences $E_1, \ldots, E_n$) through which the event characteristic could have a detrimental effect on the person or group of persons in question. The description must contain:

- the propagation of damaging components of the event characteristic, e.g. smoke in case of fire in accordance with the prevailing local conditions, M;
- the local distributions of the persons exposed to the risk, B;
- protective actions and counter-measures (evacuation, fire fighting, etc.), G.

In addition, probability estimates are needed for the different possible values of the components of M, B, and G. The arbitrarily large set of possible exposure sequences is thus approximated by a finite number of sets of discrete values (m, b, g). The probability of an exposure sequence similar to the specific local conditions, m (e.g. weather conditions during exposure), the specific exposure distribution, b, and the specific protective actions and counter-measures, g, thus becomes

$$W = w(m)\,w(b/m)\,w(g/mb) \tag{1.4}$$

In eq. (1.4) $w(b/m)$ is the conditional probability for the occurrence of b under the condition m, and $w(g/mb)$ that for the occurrence of g under the condition of the simultaneous occurrence of m and b.

(d) "Damage and risk"

In this step the relation between the intensity of the damaging effects and all damages resulting from them is described. Hence for each set of values $v =$ (category k, local conditions m, distribution of exposed persons b, emergency and counter-measures g) estimates, $x(v, a)$, of the magnitude of damage for each category a must be made.

Thus, using the estimate of the expected annual frequency

$$h(v) = h(k)\,w(m/k)\,w(b/km)\,w(g/kmb) \tag{1.5}$$

the contribution $R(v, a)$ of the set of values v to the risk number is estimated as the product of the expected magnitude of damage and its expected frequency of occurrence

$$R(v, a) = x(v, a)\,h(v) \tag{1.6}$$

The estimate of the risk number for the category of damage in question, a, is the sum of the risk contributions $R(v, a)$ taken over all considered sets of values v.

It is therefore calculated according to

$$R(a) = \sum R(v, a) = x_1 H(x_1) + x_2 H(x_2) + \ldots \tag{1.7}$$

In formula (1.7) $H(x_i)$ is the sum of frequencies of those sets of values, v, which cause an estimated magnitude of damage x_i.

1.1.4.3 Uncertainties in Risk Assessments Based on Knowledge of Details

Results of risk estimates are in part based on parameters and relations which are not exactly known. A distinction has to be made between uncertainties due to

– random variations of parameters;
– inaccurate knowledge.

Uncertainties due to random variations form part of the risk to be determined (e.g. different weather conditions).

Uncertainties of knowledge arise from inaccurate knowledge of quantities like probabilities, expected frequencies, or parameters of physical models which are fixed or considered to be so during the period of time under investigation. In addition, functional laws can often be described only approximately or an approximation is deliberately used. This is particularly true for physical laws, like for example the dependence of conductivity of heat for gases upon temperature and pressure, but also for relations described by random laws using probability distributions. Such uncertainties are contained in risk estimates based on detailed knowledge.

Variations in the result for the risk due to random variations of the parameters are inherent; they cannot be reduced by improving the methods of analysis. On the other hand, uncertainties due to inaccurate knowledge can be reduced by improving the state of knowledge.

The aforementioned uncertainties occur in all steps of a risk assessment.

1.1.4.4 Quantification of Uncertainties in Risk Estimates

Uncertainties in the estimation of risk are expressed in terms of probability distributions, i.e. by indicating ranges within which the correct value of the uncertain quantity will be found with a certain subjective probability. If uncertainties thus quantified are propagated through the different steps of the risk assessment (cf. Fig. 1.1) we obtain a probability distribution for the risk number. In this way value ranges are determined which contain the correct risk number with a certain subjective probability. This probability is called "subjective confidence level".

Hence the sets of values v from Sect. 1.1.4.2 do not merely provide a point in the magnitude of damage (x)-frequency (h)-diagram as their risk contribution, but a whole region together with the numbers $g(x, h)$ as the density of the subjective probability for the location of the pertinent contribution to risk (cf. Fig. 1.2).

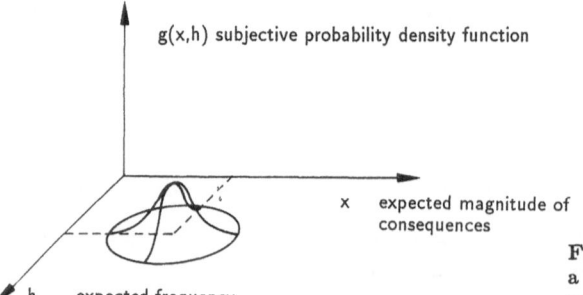

g(x,h) subjective probability density function

x   expected magnitude of
consequences

h   expected frequency

Fig. 1.2. Contribution to risk of a partial event with subjective confidence interval, from [11]

1.1.5   Formalized Representation of Risk

If risk is estimated analytically, the formalized risk definition described in detail in [12] may be used; it is based on the triplet of numbers

$$(s_i, h_i, x_i) \tag{1.8}$$

In eq. (1.8) s_i describes an event sequence which is a possible answer to the question "What can happen?", h_i is its expected frequency of occurrence and x_i represents the consequences, i.e. category and magnitude of damage. We then have for the corresponding risk

$$R_i \mathrel{\hat=} (s_i, h_i, x_i), \quad i = 1, \ldots, N \tag{1.9}$$

In eq. (1.9) N is the number of event sequences taken into account in determining the risk. Theoretically N would be arbitrarily large, since arbitrarily many event sequences are conceivable. In practice, however, event sequences are not taken into account if they are so unlikely that their occurrence may be excluded on the grounds of common sense [1].

As already explained, uncertainties are unavoidable in determining frequencies of occurrence and magnitudes of damage. They can be accounted for in relation (1.9) by writing

$$R_i \mathrel{\hat=} (s_i, g_i(x, h)), \quad i = 1, \ldots, N \tag{1.10}$$

where $g_i(x, h)$ is the subjective probability density from Sect. 1.1.4.4.

1.1.6   Presentation of Estimated Risk Numbers

The basic elements of the risk number are the magnitude of damage and the frequency with which a damage of this magnitude and of the category in question is to be expected during the time interval under consideration (mostly one year). In principle, it is sufficient to present the risk number together with the subjective confidence interval, if

– only magnitudes of damage 0 or 1 are possible (e.g. for the individual risk of the damage category "loss of human life"), or
– the magnitudes of damage per event are not too different and the event occurs frequently.

However, the requisites for the type of presentation are different if the risk numbers contain contributions from rare events with potentially large magnitudes of damage. Thus, the risk number "0.01 per year" means, for example, that, averaged over the various possibilities during one year, the magnitude of damage per year is 0.01. This number can result from

- 99 possibilities with magnitude of damage 0 and 1 possibility with magnitude of damage 1, or from

- 999 999 possibilities with magnitude of damage 0 and 1 possibility with magnitude of damage 10 000.

If the possibility to suffer damage exists only during the next 100 years, the risk number 0.01 per year states, in the first case, that damage of magnitude 1 is to be expected during this period of time. In the second case, however, it is meaningless to speak of expected damage of magnitude 1, because there will either be no damage (probability = 0.9999), or damage of magnitude 10 000 (probability = 0.0001) within these 100 years. For this reason risks from rare events with potentially large magnitudes of damage are described by both the magnitude of damage per year and its probability of occurrence (respectively, magnitude of damage per event and its expected annual frequency of occurrence). Additionally, the risk number is also given. In general, risk is stated as the probability that the magnitude of damage per year exceeds some given value X^*, respectively as the expected annual frequency of damage of a magnitude $X \geq X^*$. In the case of the frequencies all expected frequencies of risk contributions with magnitudes of damage $X \geq X^*$ have to be summed. This summation is already included in the representation of risk by means of the so-called complementary cumulative distribution function (CCDF), which is called complementary, because it gives the expected frequency for $X \geq X^*$, whereas the probability distribution gives that for $X < X^*$. Thus, for a given value X^*, the complementary probability distribution answers the question: "What is the expected annual frequency of damage of magnitude $\geq X^*$?"

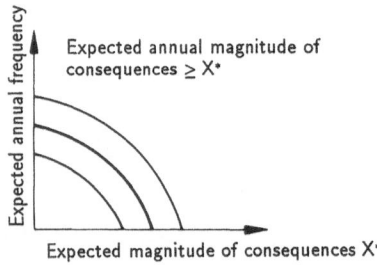

Fig. 1.3. Complementary cumulative distribution function (CCDF) with subjective confidence interval, from [11]

Figure 1.3 shows a complementary cumulative distribution function. Additionally, subjective confidence intervals are given for the different risk contributions (cf. Fig. 1.2). They form an intuitively clear band about the determined complementary probability distribution. This band is the subjective confidence interval of the curve and it states:

"The propagation of the quantified uncertainties (cf. Sects. 1.1.4.3 and 1.1.4.4) through the steps in Fig. 1.1 indicates, that the correct curve lies somewhere in the region between both limiting curves, at $P\%$ subjective confidence level, provided the influence of the non-quantified uncertainties is negligible".

An interval of analoguous meaning can also be given for the risk number.

The separate representation of the risk for the different damage categories – for example, early and late fatalities – makes it impossible to trace which magnitudes of damage of the different categories stem from a set of values v from Sect. 1.1.4.2, i.e. which of them have been caused by the same event sequence. Such relationships would have to be illustrated for the different sets of values by using tables.

1.1.7 Problems of Delineation

It is important to define clearly the scope of the risk analysis. For example, in the case of nuclear power plants, the risk resulting from normal operation and that from accidents has to be distinguished. It must be made clear whether the operation of the power plant alone or the entire fuel cycle are to be investigated. For other types of electricity generation, for example the use of solar energy, risk contributions from accidents are generally small. However, the risk resulting from the construction of the plant may be significant due to extensive construction work and the large volume of construction materials. In the case of process plants it is important to define whether the chemical process as such, or also transport and storage of the required substances are to be considered. The distinction between the risk from normal operation and that from accidents is also important for this type of plant.

A further question to be considered besides the risk resulting from the utilization of an energy source or of a chemical process could be the risk of insufficient supply resulting from its non-use.

If different technologies are to be compared, the time of reference is important. Since the risk of a technology also depends on its degree of maturity, the question arises whether all investigated systems should be judged on the basis of their technical development level at an identical point in time or if credit should be given to less developed technologies for a possible improvement of safety in the future. This is particularly important, because, in general, risk comparisons are made with regard to future decisions. These few examples may suffice to demonstrate the necessity of a neat delineation.

1.2 Potential of Probabilistic Methods for Safety and Risk Valuation

Frequently, the risk of complex technical systems cannot be derived directly from statistical observations; its estimation requires a probabilistic risk analysis, which has as its objective to assess the damages caused by accidents and their expected frequencies of occurrence.

As already explained in Sect. 1.1.4.2, such analyses are based on the identification of representative sets of conceivable accident sequences and the determination of their effects. For this purpose models of the technical systems and their components are developed and analysed. Accident sequences in plant systems are normally represented by event sequence diagrams (event trees), which in a simplified way describe the potential effects of accident-initiating events depending on the functioning or failure of the safety systems required for their control. The probabilities of failure of these systems are estimated by fault tree analyses. The use of these probabilities in the event trees then permits the estimation of the expected frequencies of occurrence of the damage-causing event sequences. Further investigations are made to determine the loads on additional safety devices and their potential failure modes; for example, the failure of containments and the quantities of dangerous substances which would then be released. The resulting damages are calculated with models describing the transport of dangerous substances and their effects on man and environment. Owing to their analytical procedure, probabilistic reliability and safety analyses account for the interactions of all important design features and operating conditions as well as the environment.

The analysis provides

- the topology of accident sequences;
- quantitative descriptions of damage-causing events and estimates of their expected frequencies of occurrence;
- the event sequences which contribute significantly to the risks;
- insights into the adequacy of plant design and operational modes by determining those plant components and modes of operation which contribute most to the expected frequencies of the risk dominating event sequences.

This provides the basis for judging

- the level of safety of a plant;
- the safety relevance of new scientific and technological results or of specific incidents during plant operation;
- promising approaches for the improvement of safety.

The insights gained from reliability and risk analyses can be used in engineering for

- eliminating weaknesses, also at interfaces between systems;
- identifying additional possibilities for improving plant safety;
- achieving a balanced design with regard to safety;
- improving modes and specifications of operation;
- improving the training of operators;
- identifying parameters with significant influence on accident consequences;
- determining key research areas and research tasks.

1.3 Overview of the Following Chapters

After giving a general description of the methods of risk analysis for technical systems in Chap. 2, the first part of Chap. 3 is devoted to peculiarities in
analysing light water reactors. In the second part the essential aspects of the
American Risk Study, WASH 1400 [13], of the German Risk Study (DRS),
(Phases A [12] and B [14]), and of NUREG-1150 [15] are presented. Further
sections of Chap. 3 deal with the risk analysis of the Sizewell plant [16] and
with some other studies conducted in the USA in the framework of the research programme initiated after the reactor accident at TMI-2.

Risk analyses for fast breeder [17] and high temperature reactors [18 – 19]
are described in Sect. 3.3. In addition to presenting the essential results, specific
aspects of the methodology compared with the analysis of light water reactors
are emphasized. The last section of Chap. 3 is devoted to risk studies for the
nuclear fuel cycle.

Risk analyses for process plants are described in Chap. 4; in particular the
Canvey Study [20] and the Rijnmond Study [21]. The process of analysis, which
differs somewhat from that used for nuclear plants, is explained. The procedure
employed in the Rijnmond Study is compared to that of the DRS-A [12].

In Chap. 5 a survey and a critical valuation of risk comparisons for electric
energy generation are presented.

Chapter 6 is devoted to possible uses of probabilistic considerations in decision making. A short description of the legal foundations and the practical
procedure for determining the safety of nuclear plants in the Federal Republic
of Germany followed by an overview of the state of probabilistic design and
safety goals in various countries concludes the presentation.

References

1. Blümel, W.; Wagner, H.: Technische Risiken und Recht. KfK 3275 (1981)
2. Lukes, R. (Hrsg.): Gefahren und Gefahrenbeurteilung im Recht Bd. I-II. Köln, Berlin,
   Bonn, München 1980
3. Bochmann, H.-P.: Gefahrenabwehr und Schadensvorsorge bei der Auslegung von Kernkraftwerken (Vortrag beim 7. Atomrechtssymposium). Göttingen 16–17. März 1983.
   Schriftenreihe Recht, Technik, Wirtschaft Bd. 31, Köln 1983
4. Renn, O.: Wahrnehmung und Akzeptanz technischer Risiken Bd. I-IV. Jül-Spez-97 (1981)
5. Rowe, W.D.: An anatomy of risk. New York, London, Sydney, Toronto 1977
6. Hartwig, S. (Hrsg.): Große technische Gefahrenpotentiale – Risikoanalysen und Sicher-
   heitsfragen –. Berlin, Heidelberg, New York 1983
7. LPG, A study – A comparative analysis of the risks inherent in the storage, transship-
   ment, transport and use of LPG and motor spirit. TNO, Apeldoorn 1983
8. Goldschmidt, L.; Beck, G.; Bill, F.: The Danish Energy Authorities' use of risk analysis.
   SRE Symposium 1987, Helsingor, Denmark, Oct. 1987
9. Risk Assessment – A Study Group Report. The Royal Society. London 1983
10. Inman, W.H.W.: Risks in medical intervention; in: M.G. Cooper (Ed.) Risk – man-made
    hazards to man. Oxford 1985

11. Kaplan, S.; Garrick, B.J.: On the quantitative definition of risk. Risk Analysis, Vol 1, No. 1 (1981) 11-27
12. Deutsche Risikostudie Kernkraftwerke. Eine Untersuchung zu dem durch Störfälle in Kernkraftwerken verursachten Risiko. Köln 1979
13. Reactor safety study – An assessment of accident risks in US commercial nuclear power plants. WASH 1400 (NUREG-75/014) 1975
14. Deutsche Risikostudie Kernkraftwerke-Phase B. Köln 1990
15. Severe accident risks: An assessment of five US nuclear power plants. NUREG-1150 (Vols 1 and 2, June 1989, Second draft for peer review)
16. Sizewell B – Probabilistic safety study. WCAP 9991 Rev. 1
17. Risikoorientierte Analyse zum SNR-300. GRS-51, Köln 1982
18. HTGR accident initiation and progression analysis status report. ERDA-Report, GA-A-13617, I-VII, General Atomic Company, 1975-1976
19. Sicherheitsstudie für HTR – Konzepte unter deutschen Standortbedingungen. Hauptband zur Phase IB, Jül-Spez-136 (1981)
20. Canvey: An investigation of potential hazards from operations in the Canvey Island, Thurrock Area. Health and Safety Executive, London 1978
21. Risk analysis of six potentially hazardous industrial objects in the Rijnmond Area – A pilot study. A report to the Rijnmond Public Authority. Dordrecht, Holland / Boston, USA / London, England 1982

2 Methods of Risk Analysis

Risk analyses are performed for technical installations which potentially cause major hazards to their environment. Such hazards usually derive from the use of substances which, through toxic or radioactive effects or the release of energy, can have a detrimental effect on health and damage property. The scope of risk analyses can vary. It can be divided into three levels [1]:

Level 1: Analysis of the operating and safety systems of the plant.
Level 2: In addition to the analysis of level 1, potential loads on containments of the plant are assessed.
Level 3: In addition to the investigations of level 2, the consequences of accidents are determined.

A level 1 study mainly consists of a reliability analysis of the operating and safety systems and of their modes of operation. Its objective is to identify accident sequences leading to potentially dangerous processes in the plant. At the same time the major causes of such sequences and their expected frequencies of occurrence are established. The investigation comprises the step "event sequences" shown in Fig. 1.1. The analyses allow one to judge whether the safety level of the technical plant systems, their design and modes of operation are adequate for avoiding dangerous processes inside the plant.

Additionally, in a level 2 analysis loads on containments caused by dangerous processes and their failure modes are determined. These are obtained from the corresponding investigations of the plant systems. Results are modes and times of failures of barriers, types and quantities of released dangerous substances or energy, and the expected frequencies of such occurrences. They provide certain insights into the risk of the plant.

Additionally, in a level 3 analysis the transport of dangerous substances and energy in the environment of the plant is investigated. Resulting consequences and their expected frequencies of occurrence are assessed. Hence, the results of the plant systems analysis are combined with the analysis of dangerous processes inside the plant up to the destruction of containments. These, in turn, are coupled with the results of accident consequence calculations. Thus, a comprehensive appraisal of the risk of a plant becomes possible.

A complete risk analysis comprises five tasks, viz.:

1. Collection of the basic plant data.
2. Identification of initiating events.
3. Event sequence and reliability analyses.
4. Investigation of the release of dangerous substances and energy.
5. Assessment of the accident consequences.

In the first task basic information on the plant, the applied procedures, and the processes involved is collected. The following three tasks concern the investigation of potentially dangerous processes inside the plant including the possibility of damage to containments or of their destruction. In the fifth task the transport of dangerous substances in the environment or the environmental impact of energy and the resulting detrimental effects are investigated. The expected frequencies of occurrence of the event sequences thus considered are estimated.

2.1   Plant Systems Analyses

2.1.1   Collection of Basic Plant Data

The basic plant data provide the fundamental information on the plant and the hazards potentially caused by its processes and operational procedures. They refer to the

– design of the plant and its systems,
– operating conditions and operational procedures,
– site of the plant and its surroundings.

In particular, data which are essential for the assessment of hazards are compiled, i.e. information on:

– dangerous substances,
– quantities of these substances and their distribution inside the plant,
– dangerous conditions of the plant,
– available safety and protection devices.

Details on the site and its surroundings are only required if a level 3 analysis is to be performed. Information about the plant comprises a description of the plant and its structure, and of the design and construction features of its systems and components. In addition, operating and handling procedures are described. For characterizing the site, information about the population distribution and the traffic situation in the vicinity of the plant and on the location of protection zones is needed.

2.1.2   Identification of Initiating Events

On the basis of the plant data, initiating events must be identified which potentially endanger the plant personnel and the population. It is useful to distinguish between plant internal and external events:

- examples of internal events are
  - mechanical failures of active components (e.g. pumps) and of passive components (e.g. pipework or vessels);
  - malfunctions or failures of measuring or control devices;
  - loss of energy and media supply;
  - human error;
- examples of external events are
  - natural phenomena like lightning, earthquakes and flooding;
  - impacts from neighbouring industrial plants;
  - impacts from means of transportation (e.g. aircraft crash or the explosion of a road tanker);
  - sabotage.

It is impossible to analyse all conceivable initiating events in detail. Rather it is sufficient to deal with those which are essential, i.e. which are significant as to frequency of occurrence or magnitude of damage or both.

The expected frequencies of initiating events are generally derived from observation. They are either estimated directly from operating experience (e.g. electrical power grid failure) or the initiating event is broken down into subevents for which operating experience is available; its frequency of occurrence is then calculated by a fault tree analysis (cf. [2, 3] and Sect. 2.1.4.2). Apart from that there are cases where the frequency of occurrence is obtained from model calculations (e.g. large breaks of pipes of a nuclear reactor coolant circuit [4]) or estimated by experts.

2.1.3   Event Sequence Analysis

Event sequence analysis is primarily used in risk studies for nuclear power plants. Starting from a defined initiating event (e.g. pipe break) and depending on the functional success or failure of the operating and safety systems required for coping with this, its different possible consequences are determined [5, 6]. The paths of events resulting therefrom are combined in an event sequence diagram, also called event tree, as shown in Fig. 2.1.

By simulating the dynamic behaviour of the plant, the systems which have to function are identified along with those which have to be activated additionally in order to control the initiating event. The simulations are based on mathematical models of the physical or chemical processes involved. Each path of the event tree is the static description of a process which is continuous. This process is represented by a few points in time (junctions) at which, depending

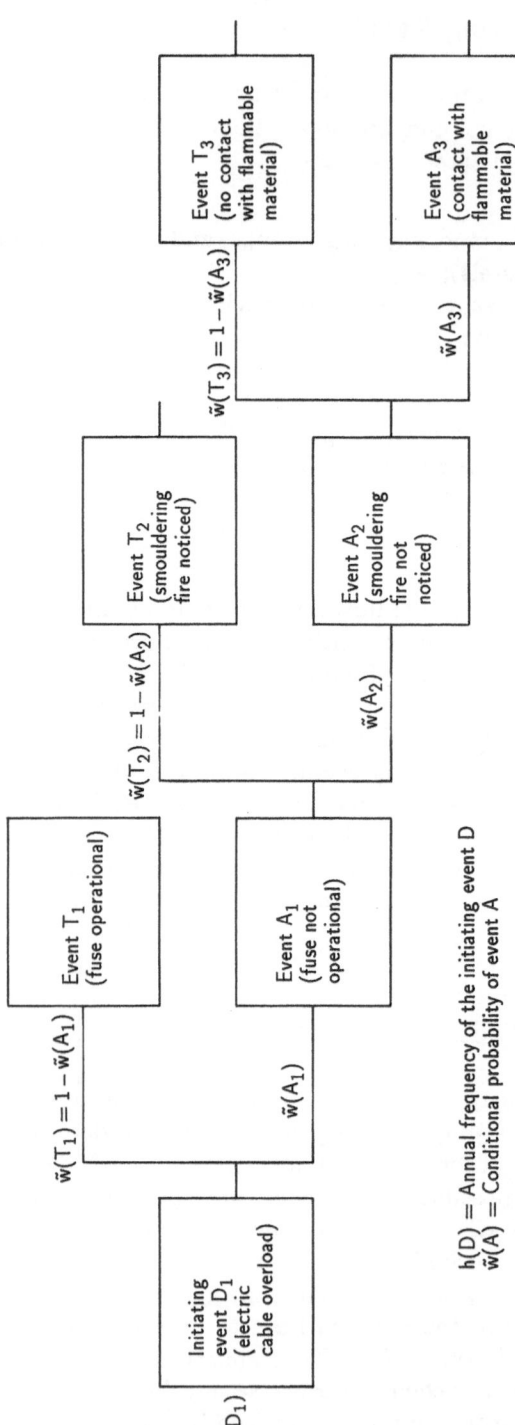

Fig. 2.1. Example of an event tree

$h(D)$ = Annual frequency of the initiating event D
$\tilde{w}(A)$ = Conditional probability of event A

on the success (upward path in the diagram) or failure (downward path in the diagram) of the required systems, its further progression is determined.

For fixing the minimum success criteria of the system functions, i.e. the minimum number of systems and functions needed for coping with an initiating event, results from accident progression simulations are frequently used which were carried out for other purposes, e.g. reactor licensing.

The event sequence analysis consists of two sub-tasks, namely

- plant systems analyses concerning those aspects of the event sequence which are determined by the action of the operating and safety systems, and
- investigations of event sequences which lead to the release of dangerous substances and energy as a consequence of the failures of the operating and safety systems of the plant.

The event trees of the first sub-task contain all paths which proved significant in the plant dynamics investigations and because of their requirements on operating and safety systems. In general, the binary logic is used, that is, systems are either considered fully operative or totally failed. Possible intermediate states are assigned to one of the two states, usually the failed state. The second sub-task is only required for an analysis of level 2 or 3.

The following aspects are relevant for an event sequence analysis:

- dependencies between different system functions may exist, because countermeasures to an initiating event are frequently performed by systems which are not independent of one another. The requirements on the system functions normally depend on the event sequence in question and on the type of initiating event.
- secondary failures (see Sect. 2.1.4.7) may occur as a consequence of previous system failures. The structure of the event sequences, that is, the chain of subsequent events, corresponds to the time history of the accident. Hence, when dealing with an event of the chain the consequences of earlier events must be considered. For example, if water leaking from a pipe renders a measuring device or a safety system inoperative, then this would have to be accounted for in further analysis.

2.1.4 Reliability Analysis

The evaluation of event trees requires numerical values for the reliabilities of the systems in order to assign probabilities to its paths. Observations which allow the reliability of the system to be estimated directly are usually not available, especially if failures are rare owing to high system reliability. However, the failure rates of components, which normally exist in larger numbers if several systems are taken as a basis, can be evaluated statistically. Parameters obtained in this way are the starting-point for estimating system reliabilities. Normally, components have several functions, for example, the opening and closing of a valve. Therefore, it must be determined which of its failure modes contributes to system failure. For this reason we often speak of the failure of a function or of a functional element instead of the failure of a component.

Besides the design of a system, its operational scheme must be taken into account in the reliability analysis. For example, emergency diesel generators are only started on demand; on the other hand, most cooling water systems are in permanent operation.

If a system function is required at a certain point in time, i.e. on demand, then the probability to be in a failed state at this instant is denominated "unavailability". If a system function must be maintained during a certain interval of time but this is not achieved, then we speak of "failure probability" or "unreliability". Often a combination of both is required, e.g. when an emergency cooling system is modelled. It must be available on demand and perform its function for a certain period of time. Unreliabilties and unavailabilities, respectively their complementary values to 1, are the probabilities assigned to the paths of the event trees (cf. Fig. 2.1).

2.1.4.1 Determination of Failure Probabilities and Unavailabilities of Components

In a probabilistic systems analysis component behaviour is described by failure probabilities and unavailabilities. Procedures and operator actions are treated like components. An independent functional element is assigned to each function of a component of the investigated system. A single functional element is also used for treating dependent failures of a certain function of several redundant components (cf. Sect. 2.1.4.7).

The failure behaviour of a functional element can be described in one of the two ways [7]

- by the failure rate λ.
  The failure rate can be interpreted as the relative decrease of the number of unfailed functional elements which occurs in a unit of time.
- by a probability of failure on demand (unavailability) p.
  The probability of failure on demand is the probability of failure of the functional element on demand, i.e. the probability that the component function has failed in the time interval prior to or at the instant of time of the demand.

Both quantities are derived from experience. They are estimated by evaluating statistically observations made during the operation of similar technical systems. Thus, mean values of the failure rate, λ, or of the unavailability, p, are calculated from the failure behaviour of several similar components working in a comparable environment. In general, both parameters are not constant in time. This is subsequently explained for the failure rate, λ, whose time history can frequently be described by a "bathtub curve", as shown in Fig. 2.2.

In the initial phase of operation so-called early failures may occur, e.g. due to latent design defects or quality or manufacturing deficiencies which have

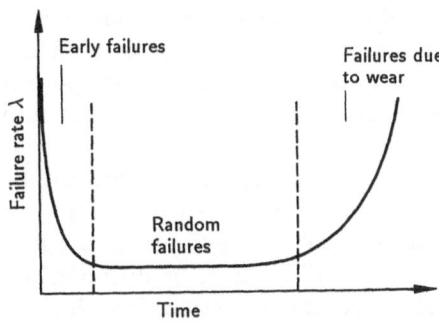

Fig 2.2. Time history of failure rates

remained undetected despite quality assurance and burn-in tests. The number of faulty components decreases with increasing time of operation due to repair or replacement until only components of the same quality level remain. Towards the end of the component's life the failure rate may increase again due to wearout and aging. During the major part of the component's life its failure behaviour is nearly random, i.e. not determined by systematic failures. Therefore failure rates are taken as constant. In this case they are the reciprocal values of the mean lifetimes of the components. The random failures are described by an exponential distribution, i.e. the failure probability $q(t)$ of a component as a function of time is given by

$$q(t) = 1 - \exp(-\lambda t) \tag{2.1}$$

The evaluation of operating experience generally provides time averaged values for the failure rates or failure probabilities. These constant values are used in the reliability analyses. They usually refer to independent failures of components. Since the body of observations of dependent failures is mostly small, specific analyses are necessary for estimating their influence, as explained in Sect. 2.1.4.7. A special role is played by operator errors. The different acts to be considered have to be analysed in order to assign failure probabilities to them (cf. Sect. 2.1.4.8). These are then used to quantify the "components" operator error in a fault tree.

Apart from the failure behaviour of components, their unavailability due to maintenance must be accounted for. Maintenance is understood here to include repair, replacement, and functional tests. From the moment of failure, respectively the start of the test, until the maintenance act is terminated the component is considered as failed. In the theoretical treatment of maintenance the following aspects play a role:

- the frequency of demands of the function, respectively, the time interval between regular functional tests (inspections) and their time staggering for failures which are not self-announcing;
- the assumption of immediate repair, as soon as the failure is noticed;
- the assignment of the "as good as new" property to the component when the repair is completed.

It may be necessary to consider several failure modes if several functions of a component are important for a system function. In such cases the fault tree analysis accounts for different failures of functional elements of a component. As an approximation, these failures are often assumed to be independent from one another.

Apart from repairs of components after a failure causing an operational disturbance, repairs of components for other reasons, e.g. failures with no immediate effect on operation, and replacements during preventive maintenance have to be accounted for. It may then be necessary to disconnect or dismount the component temporarily, in which case it cannot fulfil its function. If such maintenance work is carried out during the operation of the plant, the availability of the system function is reduced.

2.1.4.2   Fault Tree Analysis

Fault tree analysis is an established tool for analysing the reliability of complex technical systems. In such an analysis an undesired event (e.g. loss of coolant or release of dangerous substances) is defined. Then a search is made for all causes which may possibly lead to this. In general, a large variety of failure combinations of various components is obtained which make sub-systems fail. The failure of a sub-system may either directly cause the undesired event or do so in combination with failures of other sub-systems or components. The combinations are described by logical "AND" and "OR" and occasionally "NOT" gates, as defined in Fig. 2.3. The choice of the type of gate and the structure of the tree naturally reflect the underlying dynamic behaviour of the system in response to the initiating event, i.e. the time-dependent physical and chemical processes occurring in it.

In fault tree analyses complex relations within systems are described using the binary logic, which considers only the functioning or failure of components[1]. This, together with a suitable graphical presentation (cf. Fig. 1.1, where A_1 is the undesired event), permits a transparent and comprehensive treatment, even of very large technical systems. Specific problems like human error and dependent failures can be accounted for. The fault tree is the logical model of a technical process with regard to the undesired event, which is represented graphically by the symbols in Fig. 2.3.

Fault tree analysis is a complete procedure. Due to its deductive nature it yields all combinations of events leading to the undesired event, if it is consequently applied. Limitations are not inherent in the process of analysis but result from possible lack of knowledge and care by the analyst. Obviously, a fault tree analysis cannot reveal phenomena which are unknown at the time of its execution.

---

[1]   A logic with more than two states has been proposed as well [10, 11]. However, its applicability is limited by the difficulty of obtaining failure rates for intermediate states and of describing the dynamic behaviour of the system in these cases.

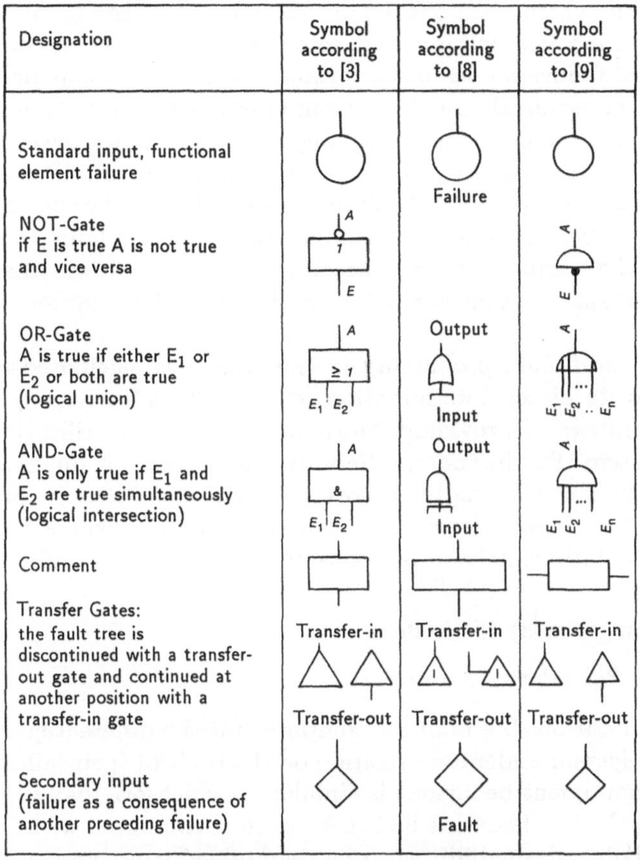

Designation	Symbol according to [3]	Symbol according to [8]	Symbol according to [9]
Standard input, functional element failure		Failure	
NOT-Gate if E is true A is not true and vice versa			
OR-Gate A is true if either E_1 or E_2 or both are true (logical union)		Output / Input	
AND-Gate A is only true if E_1 and E_2 are true simultaneously (logical intersection)		Output / Input	
Comment			
Transfer Gates: the fault tree is discontinued with a transfer-out gate and continued at another position with a transfer-in gate	Transfer-in / Transfer-out	Transfer-in / Transfer-out	Transfer-in / Transfer-out
Secondary input (failure as a consequence of another preceding failure)		Fault	

Fig. 2.3. Symbols frequently used in representing fault trees

Fault trees for complex technical systems can only be evaluated with computer programmes. Basically, two different methods are distinguished

- simulation methods and
- analytical methods.

Simulation methods may either be used for directly simulating system reliability parameters or for identifying the minimal cut sets of the fault tree. Analytical methods also serve to determine the minimal cut sets.

Minimal Cut Sets

A minimal cut set is a combination of components whose simultaneous failure is just sufficient to make the system fail. Mathematically speaking, it represents a necessary and sufficient condition for a system failure. In general, there exist several minimal cut sets for a technical system. Each of them constitutes a possible mode of its failure.

The decomposition of a fault tree into its minimal cut sets provides information about the logical structure of the system under consideration. In this way, it is possible to determine which component failures alone or in combination with others make the system fail (minimal cut sets with one or several components). The number of failure modes of the system to which a component contributes can also be determined (number of minimal cut sets in which it figures). This information is the basis for identifying weaknesses of the system; for example, minimal cut sets comprising just one component indicate the absence of redundancies. In order to assess the failure probability of the system, its structure function[2] is formed using the minimal cut sets of its fault tree. Failure probabilities of the components (e.g. according to eq. (2.1)) are then introduced into this function and thus the failure probability of the system is obtained [12].

With this procedure the failure probability of the system is calculated in two steps as opposed to its direct determination (cf. Sect. 2.1.4.3). Amongst others this has the advantage of providing information for the identification of weaknesses of the system. Furthermore, reliability parameters for different times can be calculated by substituting the corresponding probabilities (variation of t in eq. (2.1)) in the already determined structure function, whereas in direct simulation the calculations have to be repeated.

2.1.4.3 Simulative (Monte Carlo) Methods

Direct Determination of Reliability Parameters

Using random numbers, fictitious lifetimes are computed for the different components of the technical system under investigation on the basis of their failure rates. In this way the component behaviour is simulated, which originally led to the observed values of λ. The fictitious lifetimes are then compared with T, the time for which the failure probability is to be determined (mission time). All components with lifetimes shorter than the interval $[0, T]$ are considered to have failed. Components whose behaviour is characterized by unavailabilities, p, are taken as failed if the random number is smaller than p.

After this a logic function representing the fault tree is evaluated in order to determine whether or not the combination of failed components leads to a system failure. Then this process is repeated by generating a new set of random numbers, which will normally make other components fail than those which failed in the previous trial.

After a sufficient number of trials the failure probability of the system at the time T is calculated as the quotient (number of system failures):(number of trials). Other reliability parameters are determined analogously [13].

The result of the Monte Carlo simulation itself is a random variable; therefore, only confidence intervals containing the true value of the parameter can be obtained. With an increasing number of trials this interval is narrowed. The number of trials required for a specific degree of accuracy increases with the

---

[2] The structure function is a function which describes mathematically the failure of a system in terms of the failure of its components.

reciprocal value of the failure probability of the system. It can become prohibitively large for very reliable systems. A certain remedy is offered by the use of variance reducing methods; however, their application may be problematic because of the lack of rules for determining the weight factors involved.

Quantification of System Reliability Parameters by Simulative Determination of Minimal Cut Sets

Minimal cut sets are generated with the Monte Carlo method according to the scheme outlined in the previous section. Every system failure results from the failure of a set of components. This set may contain more components than are strictly required for the system to fail. These superfluous components must be eliminated from the cut set, so that the remainder constitutes a minimal cut set.

In this way reliability parameters for the system are not calculated directly; only its structure is analysed. Therefore the requirement of a large number of trials in case of reliable systems may be avoided using a fictitiously large mission time, T, which should be chosen such that the system fails in about one half of the trials [14]. This artifice makes use of the property that technical systems become increasingly unreliable with increasing operating time if there is no maintenance.

However, by Monte Carlo simulations with a practicable number of trials not all minimal cut sets of a system are found, but only those which significantly contribute to system unreliability. This property may be desirable if a system has a large number of minimal cut sets (in some cases there may be several millions), whose complete determination would be impossible because of excessive memory and computing time requirements.

2.1.4.4 Analytical Methods

In contrast to simulative methods, analytical methods find all minimal cut sets of a system. They make use of the operations of Boolean algebra and, contrary to Monte Carlo methods, require no information on component failure behaviour. This is only needed for calculating the failure probability of the system.

If the fault tree under investigation leads to a large number of minimal cut sets, a cut-off criterion must be applied in order to avoid difficulties with computer capacity. Several of these criteria exist; details can be found in [15]. Alternatively, modularization techniques, either alone or in combination with cut-off criteria, may be employed [16].

2.1.4.5 Comparison of the Methods

The method of direct simulation is a flexible tool for analysing complex systems. It allows to easily account for different maintenance strategies, limitations of repair capacities, or the activation of stand-by systems. However, results can be obtained only within certain limits of confidence. Their reduction requires a large number of trials in case of highly reliable systems. Such limitations can sometimes be overcome by employing variance reducing methods.

Procedures based on minimal cut sets give a deeper insight into the structure of the system. They provide exact solutions, or permit an estimate of the error, if cut-off is required. On the other hand, maintenance strategies can only be treated approximately or at the expense of considerable mathematical complexity. Systems represented by fault trees with a large number of minimal cut sets can originate problems with computer capacity.

Presently, there is no general rule indicating which of the methods is best for quantifying a given fault tree. Therefore, it is recommended that for practical work computer programs based on the direct simulation of reliability parameters and for the determination of minimal cut sets be available.

A comparison between two analytical and one simulative method for generating minimal cut sets made in [17] indicates that simulation is more efficient for fault trees with a large number of components and logical gates, whilst analytical methods should be used for systems with small fault trees. This result, however, depends on the type of fault tree under investigation and can therefore not be generalized.

2.1.4.6 Markov Models

Occasionally systems have to be treated in reliability analyses whose logical structure changes with time, for example if certain components or systems are activated or deactivated. Such cases can be treated by the fault tree methods presented above. In the case of an analytical evaluation this implies changing the minimal cut sets, which have to be adapted to the new logic. Thus, a renewed evaluation of the fault tree modified by the activation process becomes necessary. This may be done elegantly using the so-called phased mission models [18, 19]. Direct Monte Carlo simulation can easily handle such modifications of the system, too. Frequently, the use of Markov methods, or of semi-Markov methods, is also proposed [20, 21]. It has to be noted that in this case transition probabilities (calculated for example from component failure rates) have to be available as well and that they do not result automatically from the Markov formalism, as is sometimes erroneously asserted.

Markov processes are used to describe continuous transitions from a state X_n to a state X_{n+1} and Markov chains, if these transitions can occur only at discrete points in time. Markov processes and Markov chains are characterized by the property that the probability of the system reaching the state X_{n+1}, if it has been in state X_n before, depends only on the state X_n but not on prior states. For real systems this property is approximately satisfied if the lifetimes of the components are described by exponential distributions (hence, not in the regions of early and wearout failures), and if the underlying physical models can be described in terms of initial value problems for differential equations, which are of first order in time. This is mostly the case.

In principle, Markov processes and Markov chains allow us to quantify the probability of the system being in a certain state by solving systems of differential and difference equations, respectively. The coefficients of the systems are the transition rates. Practical implementations of the method meet considerable difficulties resulting from the large number of system states to be considered in real systems[3]. However, there are several ways of reducing the numerical effort:

- if the system can be divided into independent sub-systems, the probabilities of state of the latter can be calculated independently. Such sub-systems can then be treated using multistate analysis [10];
- under certain conditions, the number of states can be reduced by combining states into macro-states; for example, if there are symmetries in highly redundant sub-systems;
- by a suitable ordering of the states, which considers only the possible transitions, the structure of the transition matrix can be influenced in a way which favours its numerical treatment.

Even if these measures for reducing the effort are employed, Markov methods remain an extremely laborious instrument for analysing the reliability of real systems. They easily reach the capacity limits of present-day computers. In risk analyses for nuclear power plants Markov models have so far only been used for complementing fault tree analyses in some specific contexts.

2.1.4.7 Dependent Failures

Besides independent failures of functions of components, dependent failures may occur [9, 23]. Their consequences may be particularly severe if they affect redundant components or sub-systems presenting themselves simultaneously or within a short interval of time such that the failed states coexist. Such failures were called common-mode failures (CMF) in [23] and cover different types of dependent failures which are outlined below:

- failures of two or more similar or identical redundant components or sub-systems due to a single shared cause. They are referred to as common cause failures (CCF);
- failures of two or more redundant components or sub-systems resulting from a single previous failure. They are called propagating or secondary failures;
- failures of two or more redundant components or sub-systems caused by functional dependencies, i.e. resulting directly from the structure of the system. For example, functional dependencies may result from a common auxiliary system (e.g. instrument air supply), from a common control device or from human error.

In order to adequately treat dependent failures in a reliability analysis, failures of components due to functional dependencies should be modelled in the

---

[3] up to 2^N with N being the number of components of the system.

fault trees. This was done, for example in [9]. Propagating failures, as far as they cannot be excluded on grounds of spatial segregation or adequate design of the components, should also be modelled in the fault trees (e.g. secondary failures induced by missiles, by pipe whip or a humid environment).

If this is done, there remain those dependent failures which are due to shared causes (CCF) (design, construction, or maintenance errors, e.g. unsuitable lubricants in pump bearings). If possible, these should be quantified on the basis of operating experience. Several types of failures have to be distinguished in this context:

- failures which can either occur or be detected only in case of an accident,
- failures detected on demand of a function (in functional tests or in other recurrent system demands),
- self-announcing failures.

Operating experience primarily provides data for the last two types of common cause failures, which are detected during the normal operation of the plant. Failures which occur or can be detected only during an accident must usually be predicted by analytical methods. The potential for such failures will, however, only remain undetected if operational requirements or routine functional tests are not representative for the requirements on components or systems under accident conditions.

The quantification of common cause failures detected during operation or functional tests is difficult, since observations are usually scarce. This may be explained as follows:

- only a small fraction of component failures are dependent failures;
- causes of system failures which have been detected are usually eliminated. Similar failures will then only recur with an even smaller probability.

If operating experience is not sufficient for the quantification of common cause failures, recourse is taken to models. Such models are described for example in [9] and in [24, 25], where a detailed account of modelling and parameter estimation is given. In particular we have:

- coupling of failures,
- the Beta-Factor Method,
- the specialized Marshall-Olkin-Model.

The applicability and limitations of these models are summarized below.

The coupling of failures was used for the first time in [23]. The application of this method for estimating the probability of a failure of the reactor scram was sharply criticised in [26]. It is unsuitable for treating the coupling of failures of the highly redundant systems of safety and relief valves for automatic pressure relief and the limitation of pressure in nuclear reactors.

The Beta-Factor Method [27, 28] is based on the assumption that a constant fraction of all component failures (namely $\beta = 10\%$) are dependent failures. The generic β-factor of 10%, which is supposed to be valid for all types of components, was derived for the intermeshed one out of two systems formerly used in US nuclear power plants. It also contains contributions from functional dependencies and from inadequate spatial segregation. However, it does not take into account that the probability of a system failure decreases with an increasing number of redundant sub-systems or components. Likewise, the influence of administrative measures or of human redundancy (cf. Sect. 2.1.4.8) on the probability of dependent failures is not included [29].

Extensions of the Beta-Factor Model are the Multiple Greek Letter (MGL) method [24, 30] and the Multiple Dependent Fraction (MDFF) method [31]. The required parameters are determined from operating experience. Only such dependent failures are taken into account which lead to the failure of more than a given number of components at the same time. However, operating experience for the occurrence of a larger number of simultaneously failing components is very limited. Frequently, recourse must be taken to zero failure statistics. This implies the possibility of too pessimistic estimates (cf. Sect. 1.1.4.1).

With the specialized Marshall-Olkin-Model (binomial failure rate model) [24, 32, 33] not only a failure rate is estimated from operating experience but also a parameter describing the degree of coupling between redundant components or sub-systems. In this way observations of a certain number of simultaneous failures may be extrapolated to a different number of simultaneous failures. In this case sufficient operating experience must also be available. Quantifications of dependent failures using this method were carried out for emergency diesel generators, pumps, valves, and control devices based on operating experience in US nuclear power plants [34 – 37]. This quantification also included functional dependencies and propagating failures; both, as already explained, should rather be introduced directly into the fault trees, as for example in [9, 38]. The application of US data to reliability and risk analyses of different types of nuclear power plants (e.g. of German design) requires caution because of differences in the spatial arrangement, the component design, particularly of emergency diesel generators, safety valves, and control devices, and possibly in maintenance strategies. In [38] common cause failures were quantified using the binomial failure rate model. The parameters were estimated from international, German, and plant-specific operating experience. In view of the scarcity of appropriate observations the results are characterized as formula-supported estimates.

The common cause failure of a component figures as a basic event in the fault tree which coexists with its independent failure. The probability to be assigned to this basic event may be calculated with one of the described models.

Summing up, it can be stated that with the explicit modelling of functional dependencies and propagating failures, and the aforementioned models, we dispose of tools which are adequate for the treatment of the majority of dependent failures. However, successful practical applications of the models require an extension of the operating experience data base.

2.1.4.8   Human Error

When quantifying human actions in reliability and risk analyses, man is regarded as part of the system, i.e. he is treated as a system component. He has to fulfil a certain task within a given interval of time. If he does not achieve this the "component man" is considered to have failed. Man is distinct from technical components in that his behaviour is characterized by a substantially larger variability and complexity. Therefore, the description of his behaviour in terms of reliability parameters is difficult. In particular, complex interdependent actions involving several persons or decision situations are hardly amenable to probabilistic treatment. Experts on human behaviour therefore agree that only such actions or elements of actions can be described by reliability parameters which refer to skill or rule-based behaviour, as defined below.

Operator actions are classified into three categories [39], which in terms of the definitions in [40] are:

– rule-based actions (or behaviour). Behaviour in which a person follows remembered or written rules, e.g. performance of written post-diagnosis actions or calibrating an instrument or using a checklist to restore manual valves to their normal operating status after maintenance. Rule-based tasks are usually classified as step-by-step tasks unless the operators have to continually divide their attention among several such tasks without specific written cues each time they should shift attention to a different task. In the latter case, in which there is considerable reliance on memory, the overall combination may be classified as a dynamic task, especially in a post-accident condition.
– skill-based actions (or behaviour). The performance of more or less subconscious routines governed by stored patterns of behaviour, e.g. the performance of memorized immediate emergency actions following a loss-of-coolant accident or an initiating event, or the use of a hand tool by a person experienced with the tool. The distinction between skill-based actions and rule-based actions is often arbitrary, but is primarily in terms of the amount of conscious effort involved; in layman terms, the amount of "thinking" required.
– knowledge-based actions (or behaviour). Behaviour which requires one to plan one's actions based on the functional and physical properties of a system.

For the purpose of analysis actions are broken down into single elements to a degree such that reliability parameters can be assigned to these elements. The best known and most widely used method for this is THERP (Technique for Human Error Rate Prediction) [41]. This method has been in constant use in safety and reliability analyses since 1961. Subsequently it was extended and adapted to the specific situation in nuclear power plants and then used for human error quantification in [23].

The THERP method is explained in detail in [42]. It is the main procedure presented in [1] and can be divided into four steps:

1. Identification of those failure combinations which will make the system function under consideration fail, if human error is taken into account.
2. Identification and analysis of the human tasks related to the system function in question (task analysis).
3. Assignment or estimation of the relevant failure probabilities taking into account the specific conditions for the performance of the action by the so-called performance shaping factors (PSF). The basic human error probabilities of [42] are multiplied by these factors, which are greater than 1.
4. Assessment of the influence of human error on the probability of failure combinations (composed of the failure of technical components and human error) and on the unavailability of the system function (as part of the reliability analysis).

Two elements of this method deserve special attention, the task analysis and the decomposition of a complex action into its constituents. The task analysis precedes the quantitative evaluation and implies a systematic identification of the parameters which affect human reliability. In performing the decomposition existing dependencies must be identified and accounted for.

Some problems have been pointed out for which there seems to be no satisfactory solution at present using neither THERP nor any other method. These are discussed below.

As already mentioned, suitable methods and data are available for "skill-based" and "rule-based" behaviour [42]. However, their applicability to knowledge-based behaviour is limited. Objections are made primarily by cognitive psychologists. They refer to the great complexity and variability of human behaviour, particularly if decisions have to be made in new and complicated situations. It remains to be seen whether an improvement will be possible using the so-called cognitive models [43]. Other questions which still seem to be insufficiently covered are the quantification of dependencies between several subsequent steps of an action of a single operator, between joint actions of several operators, and of the control of the action of an operator by another person. The assumptions made in this context are still insufficiently supported by operating experience. Likewise the use of the so-called recovery factors for treating situations in which human error has no negative consequences because of "forgiving" properties of the system may be regarded as lacking.

In order to apply the procedures and data presented in [42] to other, e.g. German nuclear power plants, an adaptation to the specific conditions is necessary. There may be differences in the technical design (e.g. degree and extent of automation), in the ergonomic design (e.g. use of mimic diagrams), and with regard to administrative and organizational measures (e.g. organization and training of the staff), which have to be accounted for.

A short cut version of [42] is the Accident Sequence Evaluation Program (ASEP) human reliability analysis procedure [44]. It is much more of a rule-

based method than that of [42] and therefore easier to use. It is deliberately
designed to produce more conservative results than those which would be obtained applying the procedure from [42]. This is to compensate for the fact that
it is addressed to analysts not specialized in human behaviour technology.

Apart from THERP, a method similiar to the one applied in the HTGR-
AIPA Study [27] has been used in risk studies of nuclear installations. The
so-called AIPA-model relates the time available for an action, T, to the time
required for its execution, MTOR. The probability of a human error is then
quantified according to $p = \exp(-T/\text{MTOR})$. The time interval "MTOR" is
defined as the length of time required by 63% of all trained operators to complete the action successfully. It is normally determined by expert judgement.
Operating experience for its quantification is available only in few cases. The
time T needed for carrying out the action in question can mostly be derived
from plant dynamics analyses. Today the AIPA-model is only of historical interest [40]. It is meant to be supplanted by the Human Cognitive Reliability
(HCR) Model [45], which is still under development, however.

Apart from these methods, further models and procedures for describing
human error have been proposed. They are discussed in detail and compared
with one another in [40]. A short overview is given below.

Models exist for describing the behaviour of operators which do not primarily aim at its quantification; however, they are supposed to provide qualitative statements on reliability. Roughly psychology oriented models and control
theory based models can be distinguished. The following is a classification of
psychology oriented models:

– reliability models (e.g. [46, 48]),
– network models (e.g. [49, 50]),
– information processing models (e.g. [51]),
– problem-solving models (e.g. [52, 53]).

In reliability models the acting person is treated as a system component. The
"failure" of this "system component" is quantified in the same way as that of
other system components. The failure probabilities are modified, if necessary,
by performance shaping factors.

Network models can be considered as an extension of reliability models. The
system, including the necessary human actions, is represented as a network
using a kind of flow diagram. In addition to failure or success probabilities,
execution times are assigned to human actions. The parameters of the system
behaviour are then determined by Monte Carlo simulation.

Information processing models make use of the psychologists' insights into
human information processing. Information perception, properties of the memory, mental information processing, decision-making, and the action as such are
considered. Mostly, these models do not directly provide reliability data. However, they allow one to discover whether an action can be performed within
a given interval of time or with the required precision and to simulate the
consequences of mistakes.

Problem-solving models serve to model the human decision process. Only in few cases do they provide quantitative results; however, they permit an insight into human behaviour in problem solving and decision-making and hence yield qualitative statements about operator reactions.

Control theory oriented models treat the time-dependent behaviour of the man-machine-system as a closed loop system. Man is considered as a system element with certain capabilities and limitations. He performs the task of an information processing, controlling, and decision-making system element (cybernetic model of man). The models use the methods of modern control theory like stochastic processes and the description of multivariate processes by state vectors.

Attempts are made to compensate the lack of operating experience on human error by expert judgement [54]. The subjective estimation methods applied for this purpose have been in use in psychology for a long time. Most important are paired comparisons, ranking and rating, direct and indirect estimation, and multiattribute evaluation methods.

Furthermore, wider attempts are made to use operating experience and insights from simulator experiments for quantifying human error [55, 56].

Human behaviour in reliability and risk analyses has so far been quantified on the basis of a relatively coarse analysis of the tasks of the operators or by the analyst's subjective quantification of the task performance capability. A more systematic and objective evaluation of the tasks is therefore desirable. Apart from the developments of new methods described, operating experience is systematically evaluated in order to obtain data appropriate for quantifying human error. In addition, efforts are made to derive reliability parameters from the evaluation of experiments and from exercises using training simulators. In parallel with this work classification schemes were developed to make sure that the collected data are assigned coherently to the different failure categories [57].

2.1.5 Determination of Releases of Dangerous Substances or Energy

The release of dangerous substances or energy into the environment occurs only if their containments are damaged or destroyed. In general, the loads resulting from disturbed plant conditions have to be determined using mathematical models which describe complex physical and chemical phenomena. The probabilities of damage or destruction of barriers are derived from such calculations. In addition – particularly for chemical plants – spontaneous failures, i.e. failures whose causes are not or cannot be determined, of storage or transport tanks under nominal conditions must be accounted for. Expected frequencies of such release mechanisms can be derived from operating experience. Furthermore, there exists the possibility that barriers are destroyed by external events, e.g. heating up by fire or as a consequence of shock waves from explosions.

The determination of the quantities of dangerous substances released from damaged containments usually requires the use of complex models. They must account for the possibility of chemical reactions and processes like condensation, evaporation, agglomeration and deposition, which may occur during a release. The quantities of released substances and the intensity of their detrimental effects may strongly depend on the time history of these processes.

The final results of the plant systems analysis are extent, location, time history, released energy, and expected frequencies of releases from the plant which potentially cause detrimental effects. Releases occurring in different accident sequences are often combined into sets of representative releases, the so-called release categories [23] (cf. Fig. 1.1). These categories are characterized by:

- representative features for the determination of damage, e.g. quantity and type of released radioactive, toxic, flammable, or explosible substances as well as
- the sums of expected frequencies of mutually exclusive event sequences assigned to the category in question on the basis of its characteristics.

By combining event sequences to categories the volume of the required accident consequence calculations can be substantially reduced.

2.2 Determination of Accident Consequences

2.2.1 Environmental Transport

The extent of damage caused by a release of dangerous substances or energy depends, among other factors, on the type and quantity of the substances, and on their transport behaviour. Additionally, the dose-effect relationships play a role in case of radioactive or toxic substances and the combustibility or explosiveness for other dangerous materials.

The factors which influence the transport of a substance depend on the transport path. Most important are the air and water paths. However, the water path is often not considered in risk studies because its contribution to risk is usually regarded as much smaller than that of the air path (cf. [1]). In the case of atmospheric dispersion, the conditions close to the location of release (buildings, topography), the type of release (evaporation, buoyancy due to released energy), and the meteorological and topographic conditions are of importance.

During the dispersion process potential reductions of the quantity of dangerous substances by chemical reactions, radioactive decay, or dry and wet deposition have to be taken into account. For quantifying the risk, expected frequencies of meteorological situations, wind speeds, and precipitation in the surroundings of the plant have to be obtained.

If the released substances are flammable or explosible, the consequences of a release depend on whether, and at what distance from the location of release, ignitable mixtures are formed, and on whether ignition sources are present and ignition occurs. To assess the consequences, the variation of the concentration of the dangerous substance with the distance from the location of release and the probability of the presence of ignition sources at a given location have to be determined.

2.2.2 Calculation of Release Effects

In order to assess the effect of a dangerous substance the damage-causing mechanism has to be known. For example, the following categories of damage are considered: early fatalities, injuries, development of diseases (in particular cancer and genetic mutations), detrimental environmental effects and economic losses. Dose-effect relationships for toxic and radioactive materials and details of the possible exposure paths of the substances like submersion, immersion, inhalation, or ingestion are necessary for quantification. In the case of explosions or fires, the effects of pressure and heat on people and property must be known.

The significance of an event is measured by the extent of damage to people and property. Damage to humans are death, injury or diseases which break out shortly after the event or in the long run. Damage to be expected in later generations due to genetic effects has to be included as well.

Damage to people can be quantified in terms of the number of fatalities, the number of injured, or the reduction of life expectancy. In order to determine this the population distribution in the surroundings of the plant is important; in case of long-term effects of dangerous substances calculated from dose-effect relationships without a threshold, the population distribution far away from the plant is also needed. Counter-measures like seeking shelter in buildings, evacuation, medical treatment, and the interdiction of contaminated food have also to be taken into account.

Among property damage the loss of agricultural products and of natural resources, destruction, the cost of relocating the population and of decontaminating affected areas have to be taken into account. This is quantified in terms of money or areas.

2.3   Presentation of Results

2.3.1   Generalities

The results of the studies and their presentation depend on the scope and degree of detail of the analysis.

- In a level 1 analysis the following results are obtained
  - a qualitative description of the accident sequences; in addition, the causes of these sequences and, in a further step, probabilities of event tree paths;
  - the expected frequencies of sequences which potentially cause dangerous events.

The results are usually presented in tables containing the descriptions of the accident sequences and the corresponding expected frequencies of occurrence. In addition, the damage to the plant is described.

- In a level 2 analysis the description of the modes and times of failure of containments complements the results of a level 1 analysis. Additionally, the type, quantity, and time history of the release of dangerous substances and energy are provided. The results are usually presented in the form of tables, in which groups of accident sequences and their expected frequencies are pooled in release categories.
- Additionally, in a level 3 analysis the magnitude of offsite damage, and the expected frequencies of occurrence of the different types of damage are determined.

2.3.2   Collective Risks

Collective risks are usually presented by indicating the expected frequency of damage as a function of its estimated magnitude using complementary cumulative distribution functions (CCDF) (graphically or in tables). The CCDF provides the expected frequency of exceeding a given magnitude of damage. It results from the combination of many accident sequences simulated by computation. Every accident sequence consists of two basic elements:

- the plant internal accident sequence leading from the initiating event to the release. In a simple case, this sequence can be the result of a spontaneous failure of a containment, so that in effect the release coincides with the initiating event. On the other hand, the release can occur at the end of the chain of events: "initiating event → failure of control and safety systems → accident progression with overload on the containment".
- the plant external exposure sequences which comprise the transport of dangerous substances in the environment, the population distribution, the dam-

age to exposed persons and property as well as protection and counter-measures.

The results of the plant internal event sequences are the starting-point for determining the exposure sequences which are simulated with an accident consequence model. The expected frequency read off the CCDF for a given magnitude X of damage is the sum of the expected frequencies of all simulated accident sequences causing damage of magnitude X or greater. Collective risks are determined, for example, for the damage categories "early fatalities", "late fatalities", "genetic effects", or "areas requiring decontamination".

Besides the CCDF for collective risks the mean values are of interest. The mean value of collective risks of a certain type is the collective risk to be expected in the time interval of reference (mostly one year).

2.3.3 Individual Risks

In addition to collective risks, distance dependent and average individual risks are frequently also indicated.

The distance-dependent individual risk is an expected value, which is calculated by weighting the space-dependent individual damages with their expected frequencies of occurrence. The mean individual risk (expected value of individual damages) is the average individual damage to be expected within the time interval of reference. It can be interpreted as the mean value taken over all persons living at the same distance from the plant.

Individual risks are sometimes plotted as lines of equal risk (iso-risk lines) on maps of the surroundings of the plant.

2.3.4 Formalized Representation of Risk Assessment

The accident event trees (cf. Fig. 2.1) form the basis of the representation of the risk assessment process in terms of a matrix formalism [58]. This does not introduce new elements into the method of analysis but puts certain aspects into clearer perspective. In addition, the matrix formulation allows one to perform more expediently some of the mathematical operations required for proceeding from the set of initiating events to the estimation of risk.

Without entering into details, the basic idea is outlined using the diagram of Fig. 2.4. The steps of a risk analysis are treated as follows. The set of initiating events to be analysed is represented by the vector of its frequencies. This vector is multiplied by the matrices describing the response of the systems involved in the different stages of the analysis, i.e plant systems, containment, and site (representing environmental transport and health effects). The entries of the matrices are the conditional probabilities of transition between the different states, namely from the initiating events to plant states (matrix M), from plant

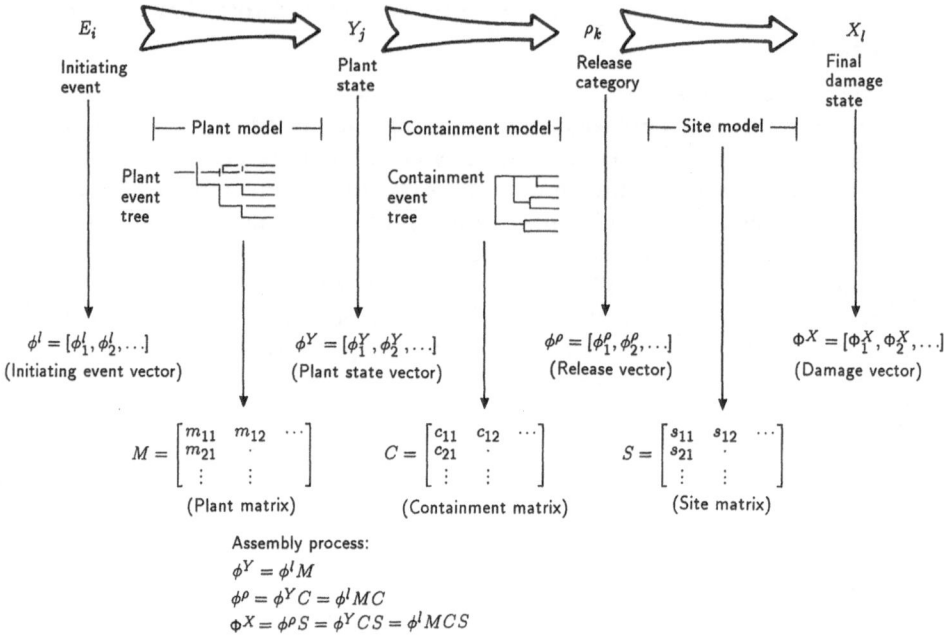

Fig. 2.4. Overview of the assembly process for nuclear plant risk analysis (from [58])

states to release categories (matrix C), and from release categories to final states (matrix S).

This formalism is now usually applied in risk analyses for nuclear power plants and was used, for example, in [38].

2.4   Uncertainty of Results

An integral part of reliability and risk analyses is the quantification of the uncertainties of the results, which comprises three tasks:

1. Determination of uncertainties in the physical and probabilistic models used in the different steps of the analysis.
   In this context three kinds of uncertainties have to be distinguished:
   - Uncertainties of the values of parameters (e.g. reliability data or coefficients in correlations) resulting from their estimation on the basis of insufficient data;
   - Modelling uncertainties resulting from the fact that models can only be an approximation of reality;

    – Uncertainties as to the completeness of the analysis owing to the impossi-
      bility to prove that all important accident sequences have been accounted
      for.
2. Propagations of the uncertainties according to 1 through the steps of the
   analysis to the final result.
3. Presentation and interpretation of the uncertainties.

The description of the uncertainties of an analysis increases the credibility
of its results. In the simplest case it is qualitative; its usefulness grows substantially, however, if the uncertainties are quantified. This quantification may
range from an indication of the upper and lower bounds of the results to probabilistically quantified subjective confidence intervals. How uncertainties are
presented depends on the objective and scope of the analysis.

Methods for quantifying uncertainties are best developed for the investigation of plant internal event sequences. They permit estimates of the uncertainties of the data for quantifying the event and fault trees and to propagate them
through the calculation. Results of the uncertainty analysis are stated as upper and lower bounds of the frequencies or as the distribution functions, from
which confidence intervals can be derived. This implies, however, that the event
tree and fault tree models are correct and complete, i.e. only the influence of
uncertainties of the data is reflected.

The uncertainties of the analysis of dangerous plant conditions, of the loads
on the containments resulting from them, and of their failure modes are larger.
In order to quantify them, recourse must frequently be taken to engineering
judgement.

In calculating offsite damages the combined influence of all uncertainties
must be taken into account, i.e. those of the plant systems analyses, the analysis of dangerous uncontrolled sequences and of the accident consequence calculations.

If CCDFs are used for presenting results, regions are provided in the
frequency-/magnitude-of-damage diagram, which contain the contribution of
a given combination of release and exposition sequences to the CCDF with a
certain confidence level. Thus, for a given level of confidence, e.g. 90%, a band is
obtained (as a global 90% confidence region) which contains the correct CCDF
with a probabilitiy of 0.9, if all unquantified uncertainties can be neglected.

For the sake of simplicity, local instead of global confidence regions are frequently determined. They are sufficient in this context. Whilst the global confidence regions are bounded by limiting lines, which contain the CCDF with a
given confidence level, the following information can be read off the local confidence regions for a fixed (therefore local) value of the magnitude of damage
or of the frequency:

– 90% confidence region of the frequency H: for a fixed magnitude of damage
  X^* there is a range of values containing the correct frequency of damage of
  magnitude X^* or greater with a probability of 0.9 (confidence level of 90%;
  vertical dotted line in Fig. 2.5).

- 90% confidence region of magnitude of damage X: for a fixed frequency H^* there is a range of values containing with a probability of 0.9 (confidence level of 90%) the correct magnitude of damage, X, which is reached or exceeded with frequency H^* (horizontal dotted line in Fig. 2.5).

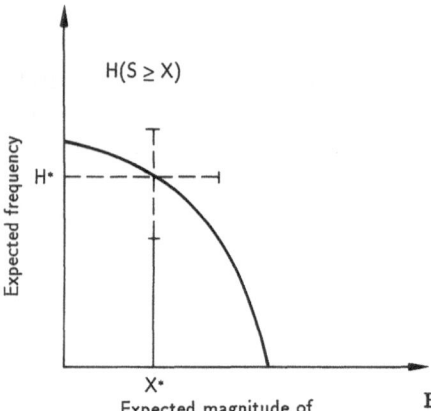

Expected magnitude of consequences X

Fig. 2.5. Location of local confidence regions for frequency and magnitude of damage

Uncertainties increase considerably on the way from plant systems analysis to offsite damage assessment. In each link of the chain of calculations "plant systems → uncontrolled dangerous sequences → loads on containments and their failure → transport of dangerous substances in the environment → effects of dangerous substances" more uncertainties are added. At the same time the uncertainties of the results in the individual steps increase in the direction towards the end of the chain. 90% confidence regions of the final results of comprehensive risk analyses may therefore cover a range of several orders of magnitude.

References

1. PRA-Procedures Guide – A guide to the performance of probabilistic risk assessments for nuclear power plants. Final Report NUREG/CR-2300 Vols. 1,2 (1983)
2. Vesely, W.E. et al.: Fault tree handbook. NUREG-0492 (1981)
3. Fehlerbaumanalyse – Methode und Bildzeichen – Teil 1. DIN-25424 (1981)
4. Schulz, H.: Comments on the probability of leakage in piping systems as used in PRAs. Nucl. Eng. Design 110 (1988) 229-232
5. Ereignisablaufanalyse – Verfahren, graphische Symbole und Auswertung. DIN-25419 (1985)
6. Deutsche Risikostudie Kernkraftwerke. Fachband 1 – Ereignisablaufanalyse. Köln 1980
7. Deutsche Risikostudie Kernkraftwerke. Fachband 3 – Zuverlässigkeitskenngrößen und Betriebserfahrungen. Köln 1980
8. IEEE Guide for general principles of reliability analysis of nuclear power generating station protection systems. IEEE Std. 352-1975, 1975
9. Deutsche Risikostudie Kernkraftwerke. Fachband 2 – Zuverlässigkeitsanalyse. Köln 1981 (English translation: Reliability analysis. Health and Safety Executive. London, July 1982)

10. Caldarola, L.: Fault tree analysis with multistate components. KfK 2761, EUR 5756e, 1979
11. Caldarola, L.: Generalized fault tree analysis combined with state analysis. KfK 2530, EUR 5754e, 1980
12. Barlow, R.E.; Proschan, F.: Statistical theory of reliability and life testing – Probability models. New York 1975
13. Kamarinopoulos, L.: Anwendung von Monte Carlo Verfahren zur Ermittlung von Zuverlässigkeitsmerkmalen technischer Systeme. IRL Bericht 14. Berlin 1976
14. Güldner, W. et al.: Computer code package RALLY for the probabilistic safety assessment of large technical systems. GRS-57, Köln 1984
15. Camarinopoulos, L.; Yllera, J.: An improved top-down algorithm combined with modularization as a highly efficient method for fault tree analysis.Reliability Engineering 11 (2) (1985) 93
16. Yllera, J.: Modularization methods for evaluating fault trees of complex technical systems. In: Kandel, A.; Avni, E. (Eds.): Engineering risk and hazard assessment. Vol. 2. Boca Raton, FL, 1988
17. Zipf, G.: Computation of minimal cut sets of fault trees: Experience with three different methods. Reliability Engineering 7 (1984) 159-167
18. Burdick, G.R.: Phased mission analysis, a review of new developments and an application. IEEE Transactions on Reliability, Vol. R-26, 1 (1977) 43-49
19. Terpstra, K.: Phased mission analysis of maintained systems – A study in reliability and risk analysis. ECN-158 (September 1984)
20. Feller, W.: An introduction to probability and its applications. New York, London, Sydney 1968
21. Gaede, K.-W.: Zuverlässigkeit – Mathematische Modelle. München, Wien 1977
22. Papazoglou, I.A.; Gyftopoulos, E.P.: Markovian reliability analysis under uncertainty with an application on the shutdown system of the Clinch River Breeder Reactor. Nuclear Science and Engineering 73 (1980) 1-18
23. Reactor safety study. An assessment of accident risks in U.S. commercial nuclear power plants. WASH-1400 (NUREG-75/014) (October 1975)
24. Fleming, K.N.; Mosleh, A.; Deremer, R.K.: A systematic procedure for the incorporation of common cause events into risk and reliability models. Nucl. Eng. Design 93 (1986) 245-273
25. Mosleh, A.; Fleming, K.N.; Parry, G.W.; Paula, H.M.; Worledge, D.H.; Rasmuson, D.M.: Procedures for treating common cause failures in safety and reliability studies. Final report. NUREG/CR-4780. Vol. 1 (February 1988), Vol. 2 (December 1988)
26. Lewis, H.W. et al.: Risk assessment review group. Report to the U.S. Nuclear Regulatory Commission NUREG/CR-0400 (September 1978)
27. HTGR Accident initiation and progression analysis status report. GA-A 13617 (October 1975)
28. Fleming, K.N.; Mosleh, A.; Kelly, A.P.,Jr.: On the analysis of dependent failures in risk assessment and reliability evaluation. Nuclear Safety 24 (1983) 637-657
29. Hörtner, H.: Problems of failure data with respect to systems reliability analysis.Nuclear Engineering and Design 71 (1982) 387-389
30. Fleming, K.N.; Kalinowski, A.M.: An extension of the Beta Factor Method to systems with high levels of redundancy. PLG-0289, Pickard, Lowe, and Garrick, August 1983
31. Stametelatos, M.G.: Improved method for evaluating common-cause-failure probabilities. Trans. Am. Nucl. Soc., 43 (1982) 474-475
32. Vesely, W.E.: Estimating common-cause-failure probabilities in reliability and risk analyses: Marshall-Olkin-Specializations. Proc. Int. Conf. Nuclear System Reliability Engineering and Risk Assessment. Gatlinburg, Tennessee, June 20-24, 1977 Edited by J.B. Fussell and G.R. Burdick, Society for Industrial and Applied mathematics, Philadelphia, 1977, 314-341
33. Atwood, C.L.: Estimators for the binomial failure rate common-cause-model. NUREG/CR-1401, EGG-EA-5112 (April 1980)
34. Atwood, C.L.; Steverson, J.A.: Common-cause fault rates for diesel generators. Estimates based on licensee event reports at U.S. Commercial Nuclear Power Plants, 1976-1978. NUREG/CR-2099, EGG-EA-5359, Rev. 1 (June 1982)

35. Atwood, C.L.: Common-cause fault rates for pumps. Estimates based on licensee event reports at U.S. Commercial Nuclear Power Plants, January 1, 1972 through September 30, 1980 NUREG/CR-2098, EGG-EA-5289 (February 1983)

36. Steverson, J.A.; Atwood, C.L: Common-cause fault rates for valves. Estimates based on licensee event reports at U.S. Commercial Nuclear Power Plants, 1976-1980. NUREG/CR-2770, EGG-EA-5485, RG (February 1983)

37. Atwood, C.L.: Common-cause fault rates for instrumentation and control assemblies. Estimates based on licensee event reports at U.S. Commercial Nuclear Power Plants, 1976-1978. NUREG/CR-2771, EGG-EA-5623 (February 1983)

38. Der Bundesminister für Forschung und Technologie (Hrsg.): Deutsche Risikostudie Kernkraftwerke – Phase B. Köln, 1990 (English Summary: German risk study nuclear power plants, phase B. GRS-74, Köln 1990)

39. Rasmussen, J.: On the structure of knowledge – a morphology of mental models in a man machine context. RISØ-M-2192, Roskilde, Denmark (1979)

40. Swain, A.D.: Comparative evaluation of methods for human reliability analysis. GRS-71, Köln 1989

41. Swain, A.D.: A method for performing a human factors reliability analysis. Monograph SCR-685, Sandia National Laboratories, Albuquerque, NM (April 1963)

42. Swain, A.D.; Guttman, H.E.: Handbook of human reliability analysis with emphasis on nuclear power plant applications. NUREG/CR-1278, SAND80-0200 RX, AN (August 1983)

43. Proc. Workshop on cognitive modeling of nuclear plant control room operators. Aug. 15-18, 1982, Oak Ridge, Tennessee. NUREG/CR-3113, ORNL/TM-8614 (December 1982)

44. Swain, A.D.: Accident sequence evaluation program human reliability analysis procedure. NUREG/CR-4772 (February 1987)

45. Hannaman, G.W. et al.: Human cognitive reliability for PRA analysis. Draft NUS-4531, NUS Corp., San Diego, CA, Rev. 3 (December 1984)

46. Payne, D.; Altman, J.W.: An index of electronic equipment operability. AIR-C-43-1/62-FR, American Institute for Research, Pittsburgh, PA (January 1962)

47. Hornyak, S.J.: Effectiveness of display subsystems measurement and prediction techniques. RADC Report TR-67-292 (September 1967)

48. Irwin, I.A. et al.: Human reliability in the performance of maintenance. Report LRP-TDR-63-218, Aerojet-General Corp., Sacramento, CA (May 1964)

49. Pritsker, A.B. et al.: SAINT: Vol. I, System analysis of an integrated network of tasks. AMRL-TR-73-126 (April 1974)

50. Blanchard, R.E. et al.: Likelihood of accomplishment scale for a sample of man-machine activities. Dunlap and Associates (June 1966)

51. Wherry, R.J.: The development of sophisticated models of man-machine system performance.In Levy, G.W. (Ed.): Symp. applied models of man-machine systems performance, AD 697939 (November 1969)

52. Newell, A.; Simon, H.A.: GPS, a program that simulates human thoughts. In: Feigenbaum, E.; Feldman, J. (Eds.): Computers and thought. New York (1963)

53. Edwards, W.L. et al.: Probabilistic information processing systems: design and evaluation. IEEE Transactions on Systems Science and Cybernetics, SSC-4-248-265 (1968)

54. Stillwell, W.G. et al.: Expert estimation of human error probabilities in nuclear power plant operations: a review of probability assessment and scaling, Washington DC, NUREG/CR-2255 (May 1982)

55. Hall, R.E. et al.: Post event human decision error: operator action tree/time reliability correlation. NUREG/CR-3010 (November 1982)

56. Sabri, Z.A.; Husseiny, A.A.: Analytical modeling of nuclear power station operator reliability. Annals of Nuclear Energy, Vol. 6 (1979)

57. Rasmussen, J. et al.: Classification system for reporting events involving human malfunctions. RISØ-M-2240, Roskilde Denmark (March 1981)

58. Kaplan, S.: Matrix theory formalism for event tree analysis: Application to nuclear risks. Risk Analysis, Vol. 2, No. 1 (1982) 9-18

3 Risk Studies for Nuclear Installations

The methods and concepts described in the preceding chapters form the basis of risk studies for nuclear installations and also for process plants. Additionally, specific methods are required which depend on the area of investigation. These are explained here for nuclear reactors taking as examples:

- the pressurized water reactor (PWR) as the most commonly used reactor type, and
- the sodium-cooled fast breeder reactor representing a more advanced type of reactor.

The explanations chiefly draw upon German designs; nevertheless in most cases they are representative of the type of reactor concerned. Special procedures applied in other areas like high temperature reactors, the nuclear fuel cycle, and process plants are discussed in the context of the respective studies.

3.1 Object and Methods of Risk Analyses for Nuclear Power Plants

The radioactive substances (fission and activation products) which are built up in the reactor core during operation constitute its radioactive inventory. This represents to a large extent the hazard potential of a nuclear power plant. The inventory is enclosed by several staggered structures, called activity barriers. In essence, the following barriers are used for confinement:

- the crystal lattice of the nuclear fuel and the fuel rod cladding,
- the pressure boundary of the primary system,
- the containment, which encloses the primary system, and
- the reinforced concrete structure which protects it, as shown in Fig. 3.1.

Nuclear power plants are equipped with safety systems designed to protect the barriers from damage with high probability. Therefore, with accidents for which the safety systems have been designed (the so-called design basis accidents: cf. Sects. 6.1 and 6.3), no offsite consequences are expected.

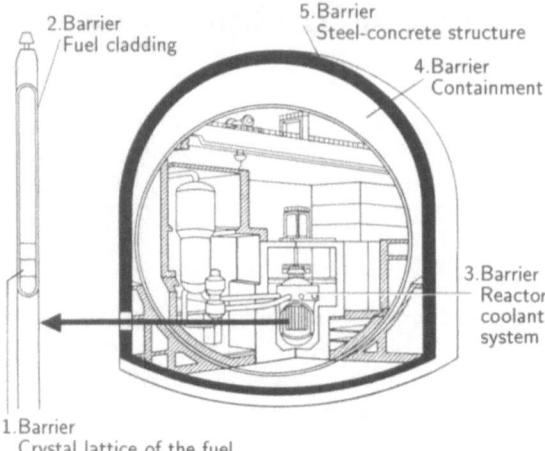

2.Barrier
Fuel cladding

5.Barrier
Steel-concrete structure

4.Barrier
Containment

3.Barrier
Reactor
coolant
system

1.Barrier
Crystal lattice of the fuel

Fig. 3.1. Activity barriers of a
pressurized water reactor (PWR)

In risk analyses the failure probabilities of safety systems as well as the modes
and frequencies of releases of radioactive substances have to be investigated.

In addition to the radioactive inventory of the reactor core radioactive materials at other locations inside the plant must be considered. Spent fuel elements
stored inside the plant are an example. However, the investigations are primarily concerned with accidents affecting the reactor core. Potential releases from
storage pools for spent fuel elements are only occasionally treated. More than
95% of the radioactive inventory of the reactor core is contained in the crystal
lattice of the nuclear fuel (in a fast breeder reactor also in that of the breeding
material). The major part of these radioactive materials can only be released
if the nuclear fuel is overheated and, in particular, if the fuel loses its crystal
lattice structure, i.e., if it melts. Therefore, the investigations concentrate on
processes which lead to overheating of the fuel.

Probabilistic risk analyses for nuclear power plants are undertaken at the
three levels outlined in Chap. 2 to a varying degree of profundity.

Investigations required for a level 1 study are discussed in Sects. 3.1.1.1 and
3.1.1.2; namely probabilistic analyses of protection and safety systems. Results of such analyses are qualitative and quantitative descriptions of the event
sequences in the reactor plant capable of causing core destruction. The set of
investigated event sequences is determined either by judicious selection of event
sequences considered important (e.g. on the basis of a coarse mesh analysis), or
determined by a systematic search for event sequences whose expected frequencies of occurrence exceed certain lower bounds. The search process is iterative
and may benefit from licensee event reports, records on operating experience,
and initiating events treated in risk studies for similar plants. The final results
of the quantitative description of such event sequences are the expected frequencies of initiation of core destruction, either listed separately for selected
sequences, or, as the sum of the expected frequencies of occurrence of all in-

dependent event sequences under consideration. In addition, the state of the plant at the onset of core degradation is described.

The study is extended to level 2 by analysing the progression of core degradation, its consequence inside the containment and the resulting containment loads. This is treated in Sects. 3.1.1.3 and 3.1.1.4. Results of such an investigation are the quantities and modes of release of radioactive substances from the containment caused by severe accidents inside the plant and their expected frequencies of occurrence.

Level 3 analysis is carried out if the atmospheric dispersion and deposition of radioactive materials and the effects on people and environment are also calculated. Results of such investigations are the magnitude of damage and the expected frequencies of occurrence for the various damage categories considered (See Sect. 3.1.2).

It is evident that only a level 3 analysis is a full probabilistic risk analysis (PRA), as defined in the previous chapters. For this reason level 1 and 2 analyses are now usually called probabilistic safety analyses (PSA).

3.1.1 Plant Systems Analysis

3.1.1.1 Initiating Events for Core Degradation or Destruction

In risk analyses, events are investigated which may lead to core degradation or destruction. The treatment of the resulting accident sequences is performed in a number of steps, as shown in Fig. 3.2.

The reactor core may be destroyed by slow melt or by a fast power excursion. Slow melt of the core is the consequence of a lasting imbalance between the heat generated in the core and that removed from the primary system; for example, through loss of the residual heat removal system which has to remove the heat still generated by the decay of the radioactive materials after a reactor scram.

A core destruction by a fast power excursion occurs if considerable reactivity is added to the reactor core quickly and in an uncontrolled way. The potential for this kind of core destruction exists if there are significant regions with positive reactivity feedback due to specific neutron physical properties. These can be encountered, for example, in sodium-cooled fast breeder reactors or in graphite moderated boiling water reactors. In the case of such a positive reactivity feedback, the power level of the reactor increases with rising core temperature.

After identifying the possible modes of core destruction all those events which could potentially initiate such a destruction have to be found. Each of these initiating events may in turn result from a variety of root causes, in which case they may be derived from fault tree analyses.

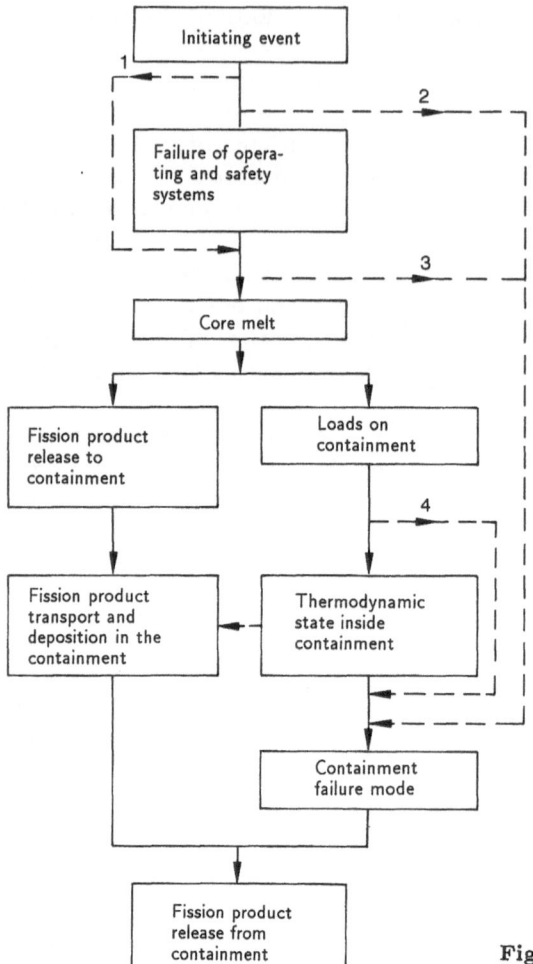

Fig. 3.2. Steps of the systems analysis for a nuclear reactor

3.1.1.2   Event Sequence and Reliability Analysis of the Protection and Safety Systems

In the event sequence and reliability analysis of the protection and safety systems the circumstances and the expected frequencies of occurrence of sequences leading from initiation of an accident to core destruction are investigated. This often includes sequences initiated by external events such as aeroplane crash, flooding, earthquakes etc.

Accident-initiating events which directly lead to core destruction, although conceivable, are extremely rare. They are therefore not taken into account in the design of the plant and cannot be controlled by the protection and safety systems (dotted path 1 in Fig. 3.2). The expected frequency of core destruction in such a case is equal to the expected frequency of the initiating event.

In many pressurized water reactors the dominant contribution to core destruction are leaks from the primary coolant system, in particular small leaks which occur either directly (e.g. pipe rupture) or in the course of a transient (e.g. from a pressurizer relief valve which remains stuck open after opening). Leaks in the primary system lead to core-melt if the active systems needed for replacing the lost coolant fail. Due to the low boiling temperature of the coolant the depressurization provokes massive evaporation and loss of coolant, finally causing core-melt. Long-term failure of the residual heat removal system leads to the opening of the primary circuit safety valves. If these cannot be closed again, this situation is equivalent to a small leak.

In sodium-cooled fast breeder reactors the failure of the reactor scram dominates all other initiating events which could potentially lead to core destruction. The reason for this is the neutron physical property of positive reactivity feedback if the coolant starts boiling. If the reactor scram fails in such a situation, a prompt supercritical condition, i.e. an extremely rapid power increase, is reached after a very short time. This may lead to the release of a significant quantity of mechanical energy. In order to avoid this type of accident, sodium-cooled fast breeder reactors are equipped with particularly reliable scram systems.

Leaks in the primary coolant systems of sodium-cooled reactors contribute little to the frequency of core-melt. Due to the high boiling temperature of the coolant which allows the system pressure to be kept low and its large heat capacity complete evaporation of the coolant is impossible if the reactor is scrammed. Loss of coolant through leakage from the reactor tank can be reliably avoided by appropriate design. Residual heat can be removed passively through natural convection in the primary and secondary systems and by the immersion coolers, even if all active coolant systems fail.

Recent risk studies for light water reactors [1, 2] have shown a considerable potential for preventive or mitigating accident management measures (denominated "recovery actions" in [1]) after a failure of the safety systems. For example, if the feed water supply to all steam generators is lost, heat removal from the reactor becomes impossible. As a consequence the primary circuit is heated up and the pressurizer relief valves will eventually open. This situation is similar to a loss-of-coolant accident with a small leak. If the safety systems and measures provided for this case should fail, core-melt would result. In this case and similar situations there is ample time before the onset of a core degradation. This time interval can be used for appropriate actions which return the plant to a safe state. If the progression of the accident towards core destruction cannot be stopped, it may still be possible to delay its onset or to mitigate its consequences.

3.1.1.3 Event Sequence Analysis for Core Destruction and Loads on the Reactor Pressure Vessel

If an accident leads to core degradation or destruction analysis of the related events is of utmost importance for understanding the progression of the acci-

dent. The initial condition for core degradation or destruction is a state where
less heat is removed than is generated in the reactor core. Reasons can be a
loss of coolant or transients.

In a light water reactor, as opposed for example to the fast breeder reactor,
recriticality during core-melt is impossible because of the low enrichment of
the fuel with U-235. Therefore, the progression of core degradation essentially
depends upon cooling conditions.

The phases of the degradation and destruction of the core of a pressurized
water reactor can be characterized as follows (see Fig. 3.3).

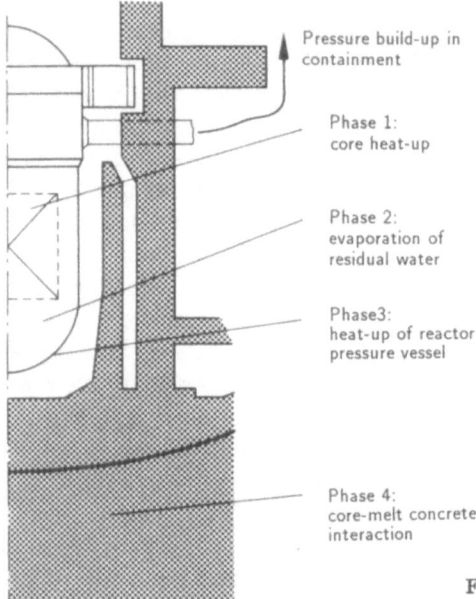

Pressure build-up in
containment

Phase 1:
core heat-up

Phase 2:
evaporation of
residual water

Phase3:
heat-up of reactor
pressure vessel

Phase 4:
core-melt concrete
interaction

Fig. 3.3. Different phases of the destruction
of the core of a PWR (from [3])

– Phase 1: Core heat-up
  The insufficiently cooled core is overheated by residual heat. Fuel rods and
  fuel begin to melt. Above 960 °C the zirconium of the fuel rod cladding
  reacts exothermally with the evaporated cooling water and accelerates the
  progression of core-melt. In this oxidation process hydrogen (H_2) is released
  from the steam and a hydrogen-vapour mixture is formed. At about 1800 °C
  Zr-UO_2 compounds may be formed whose melting point is considerably lower
  than that of UO_2. If the lower core-support plate fails due to the thermal
  and mechanical loads, large parts of the core drop into the lower plenum
  of the reactor pressure vessel. This may cause steam explosions of varying
  extent which potentially threaten the integrity of the reactor pressure vessel
  and of the containment [4].
– Phase 2: Evaporation of residual water
  The water still present in the lower plenum of the reactor vessel evaporates;
  remaining solid parts melt and form a pool of molten core material in the
  lower plenum of the reactor pressure vessel.

- Phase 3: Heat-up and failure (breach) of the reactor pressure vessel
The molten core material penetrates the wall of the reactor pressure vessel and is released to the containment together with coolant which may still be present. The distribution of the molten core material in the containment depends on the pressure inside the reactor pressure vessel at the time of its failure. If the primary system is still at high pressure, the reactor pressure vessel may be lifted off its pedestal after the failure of its bottom head and penetrate the containment or the resultant pressure build-up may be very high. In both cases the containment would be destroyed.
- Phase 4: Core-melt concrete interaction
If water is present in the cavity below the pressure vessel, it is evaporated by the molten core material thus increasing the pressure in the containment. If the cavity is dry, the core material comes into direct contact with the concrete, attacks it, expels the water bound in it, and reduces it to hydrogen due to the oxidation of zirconium, iron, and chromium. This causes an additional release of hydrogen [5]. If ignition sources are available, deflagrations or detonations of air-steam-hydrogen mixtures may occur which cause quasi-static or dynamic loads on the containment [6] (cf. Sect. 3.1.1.4) and threaten its integrity. Such an event would occur earlier than the melt-through of the concrete basemat of the containment.

With regard to the first three phases, there is agreement that the uncertainties in modelling core-degradation and core-melt phenomena are still large, but their influence on the quantification of the release of radioactive materials from the containment is relatively small [7 – 10], unless there is early failure or bypass of the containment.

The treatment of phase 4 (core-melt-concrete interaction) is based on models which have been validated and constantly improved on the basis of the experimental results, e.g. from the BETA-test facility at the Karlsruhe Nuclear Research Centre in the FRG [11].

For all the steps computer programs have been developed in several countries.

In a sodium-cooled fast breeder reactor recriticality of the melting core is possible because its fuel is highly enriched. Therefore, the progression of core destruction is determined not only by the residual heat and the cooling conditions, but also, and decisively, by possible recriticalities. These may occur after core disassembly caused by prompt supercritical reactions, but also during slow melt-down of the core.

For any further sequence of events several situations may be envisaged:

- the core melts without a release of mechanical energy,
- the molten core is retained within the reactor tank,
- the molten core penetrates the reactor tank, a renewed energy release during the melting process leads to the evaporation of sodium and other core materials.

In the last mentioned case the reactor tank and the inner containment may be damaged or even be destroyed because of very high energy releases. However, this is extremely unlikely.

The different phases of the core destruction process can be characterized as follows:

- Phase 1: initiation
  Insufficient cooling, evaporation processes, fuel rod failures, relocation of molten fuel rod cladding and of fuel inside the intact fuel element channels.
- Phase 2: transition
  Progressive melt-down and destruction of fuel elements at low power, formation of a multi-phase multi-component mixture in the core region, possibility of recriticality by fuel compaction or by fuel returning to the fission zone.
- Phase 3: integral relocation of core material and mild release
  Relocation of the multi-phase multi-component mixture in the core region, interaction of the mixture with axial and radial boundaries, release of material from the core region at low steam pressure.
- Phase 4: energetic core disassembly and release
  Extreme power excursion close to prompt criticality of the core, nuclear shutdown through relocation or ejection of molten fuel from the central core region; far-reaching core destruction and relocation of fuel inside the reactor tank.
- Phase 5: mechanical loads
  Thermal and mechanical loads on the tank internals, on the piping of the primary system, and on the tank due to the expansion of the core materials after an energetic core disassembly.
- Phase 6: residual heat removal
  Cooling of the core material after the permanent nuclear shutdown.

Uncertainties exist mainly in the detailed description of the following phenomena:

- thermal axial expansion of the fuel;
- rupture of fuel cladding in zones where the coolant has been evaporated;
- relocation of the molten core material and the fuel, dispersion of fuel by fission gas pressure and sodium and steel vapour in core regions where the coolant has been evaporated;
- fuel rod failure, relocation of the molten fuel and heat-induced reaction of fuel and sodium in fuel rod regions where the coolant has not yet or only in part been evaporated.

3.1.1.4 Containment Performance Analysis

Based on the investigations of the core destruction and the reactor vessel failure, the subsequent steps of the analysis deal with the:

- fission product releases from the primary system into the containment;

- behaviour of fission products inside the containment, and
- containment loads and the time history of thermodynamic parameters inside the containment.

The study of the release and the transport of radioactive materials inside the containment comprises the following sub-tasks:

- determination of the radioactive inventory and of the activity of the structural materials in the reactor at the time of the onset of the accident. This can be done with satisfactory accuracy using, for example, the ORIGEN-2 code [12].
- determination of the fractions of the radioactive inventory released to the primary system and to the containment during core destruction. Of great significance in this context is the treatment of chemical reactions and of deposition processes after releases from the core.

These investigations are the basis for determining the quantities of radioactive materials released to the environment, the so-called source term. The source term receives contributions from the following release processes [7 – 10]:

- fuel rod cladding-rupture,
- diffusion to the surface of the fuel,
- leach,
- fuel melt,
- core-melt-concrete interaction,
- fuel fragmentation, and
- fuel oxidation with steam.

These processes occur at different times of the core destruction.

The status of the models for the description of the individual steps is judged as follows:

- Release through rupture of fuel rods:
  The available models are not completely validated by experiments. However, the contribution to the source term from the fuel rod gap release is small. Therefore, the influence of the uncertainties is also regarded as small. For accident sequences with partial core-melt the contribution from this release mode is considered to be more significant. The application of the models requires the initial gap inventory to be known. This can be estimated from experimental observation or by using analytical models.
- Release by diffusion through the fuel:
  Diffusion is treated by classical models. Values of the diffusion coefficients are in part validated by experiments. Release by diffusion is considered to be insignificant; however, for some accident sequences it could constitute an important contribution to the source term.
- Leach release:
  Leach release is regarded as unimportant; only a few results are available.

The accident at TMI has helped focus attention on its potential contribution to the source term.

– Fuel-melt release:
Experiments have been conducted at the Nuclear Research Centre at Karlsruhe (FRG) [13] and at the Oak Ridge National Laboratory in the USA [14]. Results of the experiments are used in corresponding models. Only rough estimates of the release fractions are presently available. These estimates are extrapolations from experiments to the conditions in a reactor during core-melt.

– Core-melt-concrete interaction:
Experiments have been performed (e.g. [11]), and theoretical models are being developed to determine correlations for heat conduction, changing densities and viscosities of the molten core material, chemical reactions in the gas phase, etc. The corresponding computer codes are validated in part.

– Release through fragmentation:
Little information beyond that provided in the Reactor Safety Study (RSS) [15] is available about this release mode. The release fractions for fragmentation used in the RSS were based on measurements of radionuclide release during fuel oxidation by air.

– Fuel oxidation release:
A preliminary model is presented in [7] for the release of radionuclides from damaged fuel rods in a steam environment. It is based on the experimental observation that the rate of sintering of UO_2 is significantly greater in a steam atmosphere than in an inert or reducing atmosphere and that the release of noble gases from UO_2 is enhanced in the presence of steam.

The uncertainties existing in modelling these processes are emphasized in [10]. They will probably persist given the difficulties associated with experimental validation.

With sodium-cooled fast breeder reactors the release through fuel evaporation during core destruction by a prompt supercritical reaction is important. Considerable uncertainties exist in analytical treatment; they have to be accounted for by pessimistic assumptions.

Analysis of the release mechanisms has been a subject of intense research for many years. Knowledge about the chemical processes after release from the fuel, involving the radiologically important element iodine, has improved considerably. Accordingly, under accidental conditions in a light water reactor iodine is mainly available as water soluble cesium iodide. Its deposition behaviour differs substantially from that of organic iodine [8, 9]. However, large uncertainties still remain about potential resuspension of the already deposited fission products, about the fission product release during the core-melt-concrete interaction, the entrainment of fission products during sump water evaporation and the influence of radiation on the formation of cesium iodide. There are also additional uncertainties concerning the time history of the release of radionuclides and structural materials.

Transport, Deposition and Release Inside the Containment

Radioactive materials are released into the atmosphere of the containment as gases or aerosols. The following groups may be distinguished according to their deposition behaviour:

- noble gases,
- elementary iodine,
- organic iodine,
- aerosols.

The concentration of activity of these materials in the atmosphere of the containment can be reduced by radioactive decay and by deposition on surfaces. Reduction can be accelerated by circulation and spray systems; however, not all of these systems are available in every type of nuclear reactor.

The transport of radioactive materials in the atmosphere of the containment is analysed by computer programs which require input data about the thermal-hydraulic conditions inside the containment. These result from other computations; radionuclide transport and depletion processes, and fluid motion are therefore treated separately, although in reality they are coupled processes. Finally, the separate calculations are combined into the end result. Developments are in progress to overcome this artificial separation.

Uncertainties exist with regard to data and models used for the analysis of radioduclide behaviour. Data are inaccurate or non-existent; models treating the processes are only approximate. Other causes of uncertainty are insufficient understanding of some of the phenomena involved, and their inadequate modelling. Table 3.1 contains a survey of the most important sources of uncertainty.

Several basically different mechanisms can make the containment fail, as distinguished below:

- loads on the containment resulting directly from core destruction. Failure of the containment can be caused directly, for example, through missiles generated by mechanical failure of the reactor vessel (dotted path 4 in Fig. 3.2) or result from the pressure build-up following a core-melt accident;
- failure to isolate the containment due to the unavailability of the isolation systems (dotted path 3 in Fig. 3.2);
- certain initiating events, for example the crash of an aeroplane on a reactor building, which can directly cause the containment to fail (dotted path 2 in Fig. 3.2).

In a pressurized water reactor violent evaporation occurs in the case of leaks from the primary system due to the high system pressure and the low evaporation temperature of the coolant. In a closed containment this causes a relatively fast build-up of pressure and the corresponding loads. In sodium-cooled fast breeder reactors the pressure build-up following a loss of coolant is a slow process, because sodium is still in its liquid phase at normal reactor operating temperatures, even if the primary system were depressurized.

Table 3.1. Significant sources of uncertainty in the analysis of radionuclide behaviour (from [7])

Element of analysis	Sources of uncertainty
Inventories of radionuclides and structural materials	– No significant uncertainties
Radionuclide and structural material source term from the core	– Mode of core degradation and core-melt behaviour – Quantities of structural materials released – Chemical forms of released radionuclides – Timing of radionuclide and structural material release – Adequacy of experimental data base on releases – Validity of extrapolation of correlations based on small-scale experiments to prototypic reactor-accident conditions
Transport, deposition and release in the reactor coolant system	– Source term from the core (magnitude, physical and chemical form, timing) – Particle agglomeration – Chemical reactions – Water scrubbing of radionuclides – Thermal-hydraulic conditions – Vapour pressure of radionuclides – Validity of computer codes
Transport, deposition and release in the containment	– Source term from the reactor coolant system (magnitude, physical and chemical form, timing) – Removal of radionuclides by ice condensers [a] and boiling water reactor (BWR) suppression pools – Thermal-hydraulic conditions, particularly steam condensation on particles and hydrogen combustion – Particle agglomeration – Radionuclide attenuation during passage through containment cracks – Chemical reactions – Validity of computer codes

[a] not in German reactors

In contrast to light water reactors, sodium-cooled fast breeder reactors are designed against core destruction accidents. Therefore, the core material is retained inside the reactor tank during most accidents. A release of reactivity is only possible through leaks of the plug system. But even in the case of a damaged or destroyed reactor tank the inner containment remains as an additional barrier inside the outer containment. It provides retention capabilities including decay heat removal. Massive loads on the outer containment are only possible in the case of a failure of the inner containment. Due to the large heat capacity and heat conduction coefficient and the high boiling temperature of the coolant sodium, evaporation and build-up of loads on the containment are slow processes.

If the radioactive materials remain confined inside the containment for a long time, deposition and decay processes are important for the reduction of activity in its atmosphere. In sodium-cooled fast breeder reactors the retention capabilities of the reactor tank and of the inner containment provide an additional barrier against releases to the environment, even in the case of an open outer containment.

Besides the slow processes described, rapid events resulting from the release of large quantities of energy are investigated.

In light water reactors the following phenomena are considered:

- Pressure vessel failure at high pressure
  If the reactor pressure vessel fails while the system is at high pressure containment integrity may be threatened by the resultant pressure build-up or by missiles from the pressure vessel.
- Steam explosions
  This is an explosion-like very rapid evaporation of water coming into contact with hot molten core material. In [15, 16] a conditional probability of 0.01 was assigned to the occurrence of steam explosions with the potential to destroy the containment. More recent investigations indicate that a lower value would be more appropriate [10]. In [1] a median value of $4 \cdot 10^{-5}$ is used as a result of expert opinion.
- Hydrogen deflagration or detonation
  Great significance is attached to potential hydrogen deflagrations or detonations because they would probably destroy the containment. However, determination of the hydrogen distribution inside the containment under accident conditions is extremely inaccurate. Investigations regarding the possible extent of hydrogen deflagrations or detonations, the resulting containment loads and possible counter-measures are still in progress. The impact of hydrogen formation on containment integrity depends on the type of containment and the possible presence of inert substances, igniters or catalytic foils, which are supposed to prevent, respectively trigger hydrogen combustion before dangerous concentrations are reached.

The inner containment of sodium-cooled fast breeder reactors is inerted by nitrogen. However, if releases occur directly into the outer containment, sodium fires resulting in fast pressure build-up are possible. With extremely high mechanical energy releases the outer containment can also be destroyed, which would lead to a release of radioactivity within a short period of time.

The release of radioactive substances from the plant depends on their behaviour inside the containment and the containment failure mode. It is calculated in the last step of the investigation.

The results of the analysis of accident sequences inside the plant are the magnitude, location, time history, release of energy, and expected frequency of occurrence of radioactive releases from the plant. In order to expedite the calculations, releases resulting from various accident sequences are combined into a number of representative classes called release categories, as mentioned in Sect. 2.1.5.

3.1.2 Accident-Consequence Calculations

Accident-consequence calculations are outlined in Fig. 3.4. Initial conditions
are the modes and expected frequencies of occurrence of potential releases of
radioactivity from the containment together with the released activities of the
different nuclides. Other data required are the time elapsed from the reactor
shutdown until the beginning of the release, the time history, height and location of the release, and the quantity of thermal energy liberated. The latter
determines the buoyancy of the radioactive plume, which is formed after the
release.

For simplification in some analyses only those radionuclides are treated
which can cause significant radiation damage to the human organism; for example, iodine-131, cesium-134, cesium-137, strontium-90, and ruthenium-103
and -106.

The calculations comprise the following steps:

– atmospheric dispersion and deposition;
– calculation of doses;
– protective actions and counter-measures;
– health effects.

3.1.2.1 Atmospheric Dispersion and Deposition

Atmospheric dispersion of gases and aerosols primarily occurs in the planetary
boundary layer of the atmosphere. This layer is approximately 1000 m high.
The air flow is driven by pressure gradients and is mainly influenced by surface
friction, thermal effects due to heat taken up from the ground, and the rotation
of the earth. The lowest layer from 50 to 100 m is called the surface boundary
layer.

Atmospheric dispersion comprises several orders of magnitude of length and
time scales in the turbulence spectrum. Usually three scales are distinguished
[18]:

– Microscale: fluctuations with frequencies between fractions of seconds and
  10 minutes, and characteristic lengths between several centimeters and about
  1000 m.
– Mesoscale: time scale between 10 minutes and several hours; length scale
  between 1 km and approximately 1000 km.
– Macroscale: time scale between 1 and 10 days, and length scale between 1000
  and 10 000 km.

The global dispersion behaviour of the radioactive plume is dominated by
mesoscale and macroscale effects, whereas the local deposition behaviour is
dominated by microscale effects.

Calculations of atmospheric dispersion using the complete set of meteorological equations are extremely costly. Such analyses would provide a detailed
description of local conditions. They are not needed in risk analyses in which

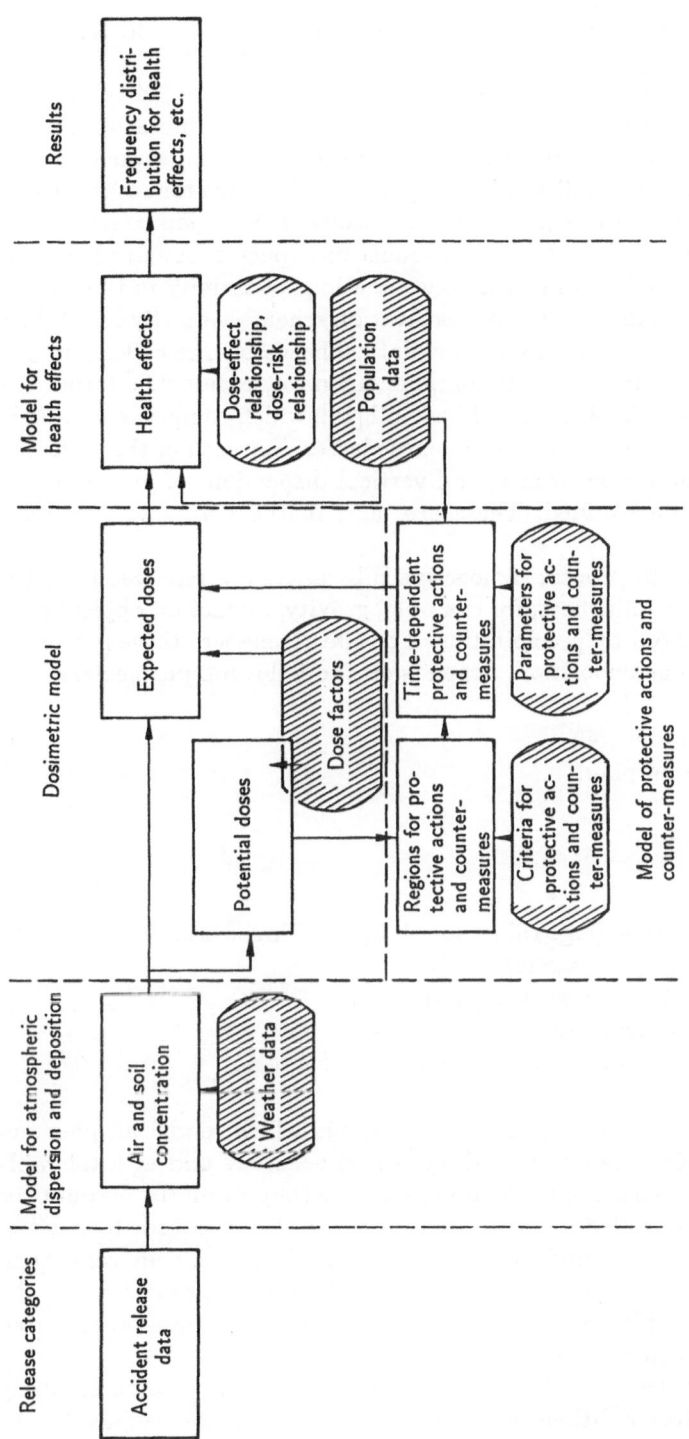

Fig. 3.4. Outline of a consequence calculation (from [17])

only global and not the site specific dispersion is considered, as in references [15 and 16].

The Gaussian dispersion model [18] is used in many risk studies to describe the vertical and horizontal activity distributions within the plume and the dependence of concentrations on the distance from the source. This model is validated by experiments up to a distance of approximately 25 km for cases without special meteorological and topographical conditions. This region roughly coincides with the area in which early fatalities must be expected. Atmospheric dispersion models serve to determine the concentrations of activity in the air and on the ground. In doing this, a simple classification scheme is mostly used which is based on readily available meteorological data. The classes are called stability categories. They characterize dispersion phenomena in parametric form. The most widely used is the classification of Pasquill [19], which comprises six stability categories corresponding to six turbulence classes. For each of these classes, parameters describing the horizontal and vertical dispersions of airborne noxious substances are determined. These are used in the Gaussian dispersion model.

During the dispersion process radioactive substances are removed from the atmosphere by various influences; for example, gravity, impact on objects close to the ground or washout by precipitation. The effectiveness of these processes, which are referred to as deposition, depends on the following parameters:

- type of vegetation and soil;
- physiological state of plants;
- topography;
- atmospheric turbulence;
- humidity;
- wind velocity;
- nature of precipitation (e.g. rain caused by a rising warm front, drizzle, distribution and size of droplets);
- solubility of aerosols in water, coagulation;
- diameter of aerosol particles;
- chemical composition and reaction behaviour of aerosols.

At present, no computer programs are available which model all phenomena according to their importance with sufficient accuracy and at a tolerable expense. Therefore, approximate methods are generally used. To account for dry deposition, the so-called source depletion model is employed. It assumes proportionality between ground deposition and the instantaneous concentration of noxious substances close to the ground without explicity describing the physical and chemical phenomena involved. Activity of the plume is reduced by the quantity deposited.

Wet deposition is described by washout coefficients; it is treated similarly to dry deposition. The conditions of dispersion are basically determined by the wind speed and direction, the state of turbulence of the atmosphere and by the occurrence or non-occurrence of precipitation, i.e. by the general time and space dependent weather history.

In modelling dispersion phenomena we have to differentiate between risk analyses performed for a collective of nuclear power plants at different sites and those for one nuclear power plant at a specific site. In the first case dispersion models are chosen that describe the phenomena correctly on the average; in the second case the meteorological and topographical data of the site considered have to be accounted for.

Data about observed weather histories are usually available in hourly cycles. For the sake of expediency samples have to be taken from these data. Special sampling techniques guarantee that the daily and seasonal cycles are sufficiently represented. These samples of weather histories are then assigned to the already mentioned Pasquill stability categories.

To determine early fatalities the spatial distribution of radioactive material in the vicinity of the plant is important; on the other hand, late health effects are primarily determined by the total quantity of released radioactive material. The atmospheric dispersion model and its parameters have little influence on the calculated late health effects.

3.1.2.2 Determination of Radiation Doses

As shown in Fig. 3.4, determination of the radiation dose is strongly linked with the model for protection and counter-measures. Some of the measures considered in this model, such as

- evacuation,
- decontamination,
- restrictions on the consumption of agricultural products,

are based on radiation dose criteria, that is, such measures are taken when certain radiation dose thresholds may be exceeded. This requires prediction of "potential" radiation doses resulting from staying permanently outdoors and from the exclusive consumption of agricultural products from the contaminated areas.

Taking into account the mitigating effects of the counter-measures assumed in the model, the expected radiation doses are assessed in the subsequent step. Energy doses for specific organs and for the whole body are calculated to characterize radiation exposure. The energy dose is the measure used for the radiation energy absorbed per unit organ mass. Its unit is 1 Gray $(Gy)^1$.

The ways by which the radiation of the released radioactive materials reaches humans are called exposure pathways. Principally, pathways resulting from release of radioactive materials into the ground and from atmospheric release have to be treated. However, releases resulting from the penetration of the

---

1   1 Gray (Gy) = 1 J/kg. The equivalent dose which takes into account the biological effectiveness of the various types of radiation is derived from this unit; it is measured in Sievert (Sv). Formerly the following units were used: 1 rad = 0.01 Gy and 1 rem = 0.01 Sv.

basemat of the containment are often not determined, because their impact on the environment is considered small compared to that of atmospheric releases. Essentially, the following exposure pathways are investigated:

- external irradiation due to the passing radioactive plume (cloudshine);
- external irradiation from radioactivity deposited on the ground (ground-shine);
- internal irradiation from radioactive substances incorporated with inhaled air, viz.
  - inhalation of airborne radionuclides from the passing radioactive plume,
  - inhalation of resuspended radionuclides which had already been deposited on the ground,
  - internal irradiation due to radioactivity incorporated with food (ingestion).

In general, the calculation of radiation exposure is performed for the following:

- bone marrow,
- bone surface,
- breast,
- lung,
- thyroid.

Injury to other organs is calculated from the whole body radiation exposure.

In order to assess genetically significant effects, radiation exposure of the testes and the ovaries is used.

Calculation comprises the determination of potential doses to the:

- bone marrow due to external irradiation by radionuclides deposited on the ground during the first few days for early health effects;
- whole body from external irradiation from deposited radioactive substances over many years for late health effects;
- whole body, bone marrow and thyroid due to internal irradiation resulting from activity incorporated with food accumulated over 50 years for late health effects.

Taking into account the protection and counter-measures which reduce the effects of irradiation the following expected doses are determined:

For early health effects

- the short-time bone marrow dose composed of doses from
  - external irradiation from the passing plume (cloudshine),
  - external irradiation from contaminated ground (groundshine),
  - internal irradiation due to inhaled radioactivity from the passing plume.

For late somatic effects

- doses for the organs
  - bone marrow,

- bone surface,
- lung,
- thyroid,
- breast,
- other tissues.

The doses from all exposure pathways resulting from the release of radioactive materials into the atmosphere are added up. Radiation exposure of the directly affected population is considered as well as the radiation effects on their descendents.

3.1.2.3 Determination of Radiation Health Effects

The absorption of radiation energy by a living cell or tissue initiates a chain of chemical, physical and biological reactions. At the end of this chain the health of the irradiated person may be injured, or, in case of radiation to the gonads (testes, ovaries), genetic damage to the person's descendents may occur. Figure 3.5 illustrates the reaction chain of biological effects of irradiation in an organism and the potential types of radiation health effects. Since both cell and organism have highly effective mechanisms to repair or eliminate the primary biological effects of irradiation, the absorption of radiation by an organ or tissue of the body does not necessarily result in the manifestation of radiation injury, except in the case of very high doses. However, there is a non-zero probability for radiation health effects. It is called radiation health effect risk and denotes the probability that health damage will follow from irradiation.

According to Fig. 3.5 we distinguish four types of radiation health effects:

- acute or early health effects which appear shortly after irradiation (e.g. acute radiation sickness);
- non-malignant late damage, manifesting itself many years after the irradiation in the affected person (e.g. fibrotic tissue changes, clouding of the retina, reduction in fertility);
- malignant late health effects manifesting themselves in the irradiated tissue only after a latent period lasting from years to decades (e.g. leukaemia, due to the irradiation of the bone marrow or tumors in other body tissues);
- genetic effects resulting from specific mutations of reproductive cells after irradiation of the gonads (testes, ovaries) causing changes in the offspring of the irradiated individual (e.g. skeletal anomalies, mental retardation, changes in eye colour, hereditary diseases).

The first three types of injury are somatic health effects, since they occur in the affected individual. This also includes injuries which can occur in later generations as a result of radiation to the womb.

The carcinogenic and genetic effects are stochastic, because only their probability of occurrence, but not the level of injury, depends on the radiation dose. This is in contrast to the non-stochastic early and the non-carcinogenic late health effects, where the level of injury depends on the dose.

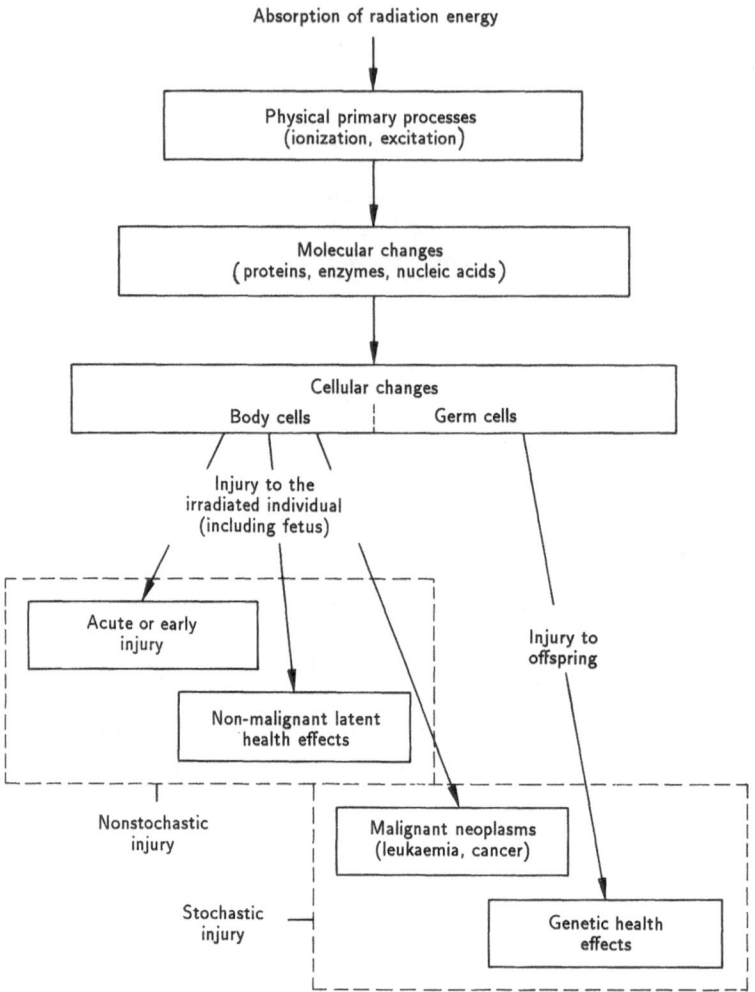

Fig. 3.5. Reaction chain of biological radiation effects and types of radiation injury (from [17])

The most important aspect in the evaluation of health damage caused by ionizing radiation is the risk in all types of injury which lead to a radiation-induced loss of lifetime (mortality risk). These are severe acute radiation sickness caused by high doses to the blood-forming organs, and malignancies like lung cancer or leukaemia.

Analysis of potential adverse effects of irradiation requires knowledge of the relation between the dose incurred by a tissue and the probability of occurrence of damage to this tissue. There is a fundamental difference between the form of the dose-effect or dose-risk relationships for stochastic health effects (cancer, genetic injury), and for non-stochastic health effects (acute injury, non-carcinogenic late health effects).

For non-stochastic health effects a threshold dose exists. As long as this threshold is not exceeded, there will be no detriment to health. Early health effects are, therefore, in general only to be expected in the vicinity of the location of release. On the other hand, experience with carcinogenic and genetic radiation effects suggests that there is no such threshold dose. They can also occur in the case of low doses. However, the probability of their occurrence decreases as the dose decreases. Carcinogenic radiation effects are predicted wherever transported radioactivity leads to radiation exposure of the population; their occurrence is far-reaching and not restricted to the vicinity of the location of release as in the case of early injuries. Some researchers also claim the existence of a threshold dose for this type of radiation health effect [20, 21]. With short-time whole body irradiation with doses below 1 Gy the resulting health damage is practically determined by the risk of carcinogenic or genetic radiation effects. For very high doses non-stochastic effects are significant with regard to reduction of the expectation of life.

The shape of the dose-risk relationship depends not only on the quantity of absorbed radiation energy, i.e. on the energy dose to the particular tissue, but also on the chronological distribution of the received dose and the type or quality of the radiation. The latter is characterized by the local density distribution of ionization in the track of the charged particles. Ionizing particles of high ionization density along the particle path, i.e. high linear energy transfer (LET), generally have a higher biological effectiveness at the same energy dose than low-density ionizing particles, i.e. particles with low LET. Among the former are α-particles, whereas X-rays and γ-rays, as well as electrons or β-rays, belong to the low-density ionizing types.

The radionuclides released during accidents in nuclear power plants with light water reactors primarily emit β- and γ-rays. In nuclear power plants with fast breeder reactors the α-emitters are also important; therefore, special attention has to be paid to the exposure of the lungs to this type of radiation.

After a reactor accident a substantial number of early fatalities can only occur

- if large quantities of radioactive materials are released to the environment during the early phase of the accident due to the failure of the containment or of its isolation systems,
- if the wind blows in the direction of a densely populated area, and
- if rain leads to high concentrations of radioactive substances close to the ground.

Early fatalities are usually restricted to the surroundings of the plant, whereas late fatalities are to be expected wherever dispersion of the released substances leads to radioactive exposure if a dose-risk relationship without a threshold is used for calculation.

3.2 Studies for Light Water Reactors

3.2.1 The US Reactor Safety Study

The US Reactor Safety Study (RSS) [15] was the first comprehensive investigation of the risk of accidents in nuclear power plants in which probabilistic methods were applied. Its objective was to assess realistically by extrapolating from the results for two reference plants the risk from potential accidents in the 100 commercial nuclear power plants which were expected to be in operation in the US by 1980. The estimated risks were compared with other natural and man-made risks. The study comprised all the steps described in Chap. 2, i.e. from the identification of initiating events to the determination of health effects from releases of radioactive materials.

Starting from initiating events, event trees were developed for a number of accident sequences under the assumption that the system functions fail if the minimum success criteria used in the licensing procedure are not satisfied, i.e. fewer systems are available than are required there. If an accident sequence leads to damage of the reactor core, complete melt-down of the core was assumed. The reference plants used were the boiling water reactor Peach Bottom with a power of 1065 MWe and the pressurized water reactor, Surry 1, with a power of 775 MWe.

Discussions about the results of this study led to the formation by the US Nuclear Regulatory Commission (NRC) of the risk assessment review group under the chairmanship of H.W. Lewis [22]. Its mandate was to demonstrate the achievements and limitations of the study, to investigate the status of the methods for risk assessment and to judge whether risk studies should be used in the licensing procedure.

The review group concluded that the reactor safety study marked a significant progress over earlier attempts in assessing the risks of nuclear energy. However, the group was unable to determine, whether the calculated core-melt frequencies were too low or too high. The uncertainty bands for the results presented in the RSS were believed to be too narrow. The fault tree and event tree analyses, combined with a suitable data base, were considered to be the best available method for the quantification of the frequencies of occurrence of accidents. Risk analyses were regarded as sufficiently complete if initiating events, accident sequences and component failures which have not been acccounted for do not provide significant contributions to risk. In this context, the review group considered such circumstances as significant which may influence the risk by a factor of 2 or more. The RSS was the first to point out the importance of accident consequences other than early fatalities. Significant improvements to the accident-consequence model and the knowledge of its sensitivity towards variations of input parameters were considered necessary before it could be used in the licensing procedure; critical parameters being, for example, the thermal lift of the plume of released substances, the deposition velocities, the effectiveness of evacuation procedures and the dose-effect relationships for early and late fatalities.

Some results of the study are presented in Table 3.8 and Figs. 3.9, 3.10 and 3.11.

3.2.2 Further Studies in the USA

After the completion of the RSS the Nuclear Regulatory Commission initiated the gradual use of probabilistic risk analyses in its regulatory process. The accident at the Three Mile Island (TMI) nuclear power station in 1979 accelerated the development of probabilistic methods and their application to the safety evaluation of nuclear power plants. As an immediate action the Reactor Safety Study Methodology Programme (RSSMAP) was initiated. Within its framework one reference plant of each of the four US vendors of nuclear power stations was investigated on the basis of the methodology and the results of the RSS. Due to design features different from those of the reference plants of the RSS the resultant expected frequencies for core-melt differed from those obtained there. The objective of the RSSMAP Studies was to identify risk-dominant event sequences and to compare them with those found in the RSS. The bases of the studies were the safety reports and additional information provided by the vendors and the utilities. Accident-consequence calculations were not performed. Reliability parameters for components and for the quantification of operator actions were taken from the RSS. Only if the design of the systems differed from those considered in the RSS were the system unavailabilities determined.

Based on results of the investigations, contributions from several accident sequences to core-melt could be reduced by modifying the systems or operating instructions.

At the time of the completion of the RSSMAP Programme the NRC considered it premature to perform risk analyses for all nuclear power plants. Therefore the Interim Reliability Evaluation Programme (IREP) was initiated. It was primarily devised to improve the probabilistic methods. Under this programme five plants were analysed and their core-melt frequencies were assessed in 1981 and 1982. In doing this, realistic system success criteria were used instead of the conservative assumptions of the licensing procedure. Offsite consequences were not addressed. Uncertainties of the results were discussed verbally, but not quantified. The IREP Procedures and Schedules Guide [23] was written.

Parallel with the IREP Programme the NRC suggested that the utilities themselves perform risk studies of their nuclear power plants. Analyses were thus carried out for plants requiring licensing decisions. In the framework of these activities a comprehensive handbook, the PRA Procedures Guide, was prepared [7] describing the methods of probabilistic risk analyses. It is to serve as a guide for the performance of probabilistic risk analyses.

The risk study programmes led to the National Reliability Evaluation Programme (NREP). Within its framework, risk studies for numerous nuclear power plants were performed using the methods described in the IREP and the PRA Procedures Guides. In addition, the NRC published the NREP Handbook [24]. In contrast to the generic character of the RSS these studies are plant-specific. Not all of them comprise all the steps described in Chap. 2. In some cases accident consequences were not assessed or only the expected frequencies of core-melt were determined. In general, the studies rely heavily on the results of the RSS.

In the NREP Programme a cost effective application of risk analyses in the framework of licensing procedures was attempted. A close relationship exists with the probabilistic safety goals, which are discussed in Sect. 6.6.

Table 3.2 gives a survey of the different programmes. Tables 3.3 and 3.4 provide details on the different studies.

Table 3.3. Scope and results of risk studies [26]

Plant	Scope of the study [a]	Results [b] (annual core-melt frequency)
Arkansas One-1	S	$5 \cdot 10^{-5}$
Big Rock Point	S,R	$1 \cdot 10^{-3}$, after impr. $9 \cdot 10^{-5}$
Browns Ferry-1	S,D	$2 \cdot 10^{-4}$, EF 5.6 [c]
Calvert Cliffs-2	S,C	$2 \cdot 10^{-3}$
Crystal River-3	S	$3.7 \cdot 10^{-4}$
Grand Gulf-1	S,C	$4 \cdot 10^{-5}$
Indian Point-2	S,R,E,D	Median: $4 \cdot 10^{-4}$, EV $4.7 \cdot 10^{-4}$ 90% centile: $1 \cdot 10^{-3}$
Indian Point-3	S,R,E,D	Median: $9 \cdot 10^{-5}$, EV $1.9 \cdot 10^{-4}$ 90% centile: $5.5 \cdot 10^{-4}$
Limerick-1,2	S,(C),R,D	$1.5 \cdot 10^{-5}$
Millstone-1	S	$3 \cdot 10^{-4}$
Oconee-4	S,C	$8 \cdot 10^{-5}$
Peach Bottom-2,3	S,C,R,E,D	Median: $3 \cdot 10^{-5}$, EF $(0.7–17.8) \cdot 10^{-5}$
Sequoyah-1	S,C	$6 \cdot 10^{-5}$
Surry-1,2	S,C,R,E,D	Median: $5 \cdot 10^{-5}$, EF $(1.4–96) \cdot 10^{-5}$
Zion-1,2	S,C,R,E,D	Median: $4 \cdot 10^{-5}$

[a] S = core-melt frequency; C = containment performance; R = risk; E = external events; D = confidence intervals calculated; EF = error factor or range; EV = expected value.
[b] Results are point values, if there is no indication to the contrary.
[c] Error factor: square root of the quotient of the upper and lower bounds of the confidence interval.

Table 3.2. Survey of the risk study programmes in the USA [26]

Power station	Type (vendor)*	Contain-ment	Power in MWe	Year of commissioning	Site category [a]	Publication	Programme [b] (analysing team)
Arkansas One I	PWR (B & W)	Dry	858	1974	II	12/81	IREP II (SNL/SAI)
Big Rock Point	BWR (GE)		75	1962	I	3/81	Utility (SAI)
Browns Ferry I	BWR 4 (GE)	MK I	1067	1973	II	7/81	IREP II (EG & G, EI)
Browns Ferry II, III	BWR 4(GE)	MK II	1067	1974/76	II		IREP II (SNL/SAI)
Calvert Cliffs I	PWR (CE)	Dry	845	1975	II	5/82	RSSMAP (SNL)
Calvert Cliffs II	PWR (CE)	Dry	880	1977	II	12/82	IREP I (SNL/SAI)
Crystal River III	PWR (B & W)	Dry	875	1977	I	10/82	RSSMAP (SNL)
Grand Gulf I	BWR 4 (GE)	MK III	1250	1982	I		
GESSAR II	BWR 4 (GE)	MK III	Planning risk study during design phase				Utility (PLG)
Indian Point II	PWR 4 (West)		873	1973	V	3/82	NREP-prel. study (PLG)
Indian Point III	PWR 4 (West)		965	1975	V	3/82	Utility (SAI/GE)
Limerick I, II	BWR 4 (GE)	MK II	1055	1985/87	V		Utility (PLG)
Midland II	PWR (B & W)		805	1983	III		IREP II (SNL/SAI)
Millstone I	BWR 3 (GE)	MK I	652	1971	IV		NREP (planned)
Millstone III	PWR (West)		1156	1986	IV		Utility (PLG)
Oconee I, II, IV	PWR (B & W)	Dry	860	1973/74	II	10/81	RSSMAP (SNL)
Oconee III	PWR (B & W)		860	1974	II	5/81	Utility (PLG)
Oyster Creek I	BWR 2 (GE)	MK I	620	1969	II	unpubl.	RSS (MIT/NRC)
Peach Bottom II, III	BWR 4 (GE)	MK I	1098	1973/74	III	75	RSSMAP (SNL)
Sequoyah I	PWR 4 (We.)	MK II	1148	1981/82	III	4/81	RSSMAP (SNL)
Shoreham I, II	BWR 4 (GE)	MK II	820	1983	IV		Utility (PLG)
Surry I, II	PWR 4 (West)		824	1972/73	II	75	RSS (MIT/NRS)
Susquehanna I, II	BWR 4 (GE)	MK II	1100	1983/84	III		Utility (NUS)
Yankee Rowe	PWR (West)		175	1961	I		Utility (EI)
Zion I, II	PWR 4 (West)	Dry	1100	1973/74	V		Utility (PLG)

[a] Population density: Group I = below average. Group II = average. Group III = slightly above average. Group IV = above average. Group V = markedly above average.

[b] Studies by the utility were mostly performed on the basis of the PRA Procedure Guide.

* B & W: Babcock and Wilcox; CE: Combustion Engineering; EG & G: EG and G, Inc.; EI: Energy, Inc.; GE: General Electric; MIT: Massachusetts Institute of Technology; NUS: Nus Corp.; PLG: Pickard, Lowe & Garrick; SAI: Science Applications, Inc.; SNL: Sandia National Laboratory; West: Westinghouse.

Table 3.4. Estimated frequencies for core-melt from published probabilistic analyses [26]

Plant and source of information	Expected annual frequency for core-melt	Plant and source of information	Expected annual frequency for core-melt
1. Arkansas Nuclear One (ANO-IREP)	$5 \cdot 10^{-5}$	7. Oconee-3 (O-RSSMAP)	$8 \cdot 10^{-5}$
2. Browns Ferry-1 (B.F. – IREP)	$2 \cdot 10^{-4}$	8. Sequoyah-1 (SEQ-RSSMAP)	$6 \cdot 10^{-5}$
3. Calvert Cliffs-2 (C.C. RSSMAP)	$2 \cdot 10^{-3}$ [a]	9. Surry-1 (RSS)	$6 \cdot 10^{-5}$
4. Crystal River-3 (Crystal River IREP)	$4 \cdot 10^{-4}$	10. Zion-1,2 (ZPSS)	$4 \cdot 10^{-5}$ [b]
5. Grand Gulf-1 (Grand Gulf RSSMAP)	$4 \cdot 10^{-5}$	11. Limerick-1,2 (LGS PRA)	$1 \cdot 10^{-5}$ [c]
6. Millstone-1 (Millstone IREP)	$3 \cdot 10^{-4}$	12. Peach Bottom-2 (RSS)	$3 \cdot 10^{-5}$

[a] Does not account for substantial modifications of the auxiliary feed water system.
[b] Total core-melt frequency (internal events only) according to Zion PRA, Vol. 10, Sect. 8.
[c] From the Limerick study version of April 1982.
Sources: "Catalog of PRA Dominant Accident Sequence Information (Draft)", EG&G Idaho, Inc., June 1983. "Interim Report on Accident Sequence Likelihood Reassessment", Sandia National Laboratories, August 1983

3.2.3 NUREG-1150

By the mid-1980s a series of new computational models for analysing the physical processes in severe accidents had been developed, among them the Source Term Code Package (STPC) [25], and the PRA Procedures Guide had been published.

Using the new techniques and the procedural framework, the risks of severe accidents in five nuclear power stations were reassessed. The results were published in NUREG-1150 [1]. This report was issued as a draft for public comment in February 1987 and is now available as a second draft, which accounts for the comments received. It reflects the improvements in methods, in the design and operation of the plants studied, and the extended information base on severe accident phenomenology.

The objectives of this study are:

- to provide a snapshot of risks reflecting plant design and operational characteristics, related failure data and severe accident phenomenological information available as of March 1988;
- to update the estimates of the RSS;
- to include quantitative estimates of risk uncertainty in response to a principal criticism of the RSS;

- to identify plant-specific risk vulnerabilities for the five studied plants, supporting the development of the NRC's individual plant examination (IPE) process;
- to summarize the perspectives gained in performing these analyses, with respect to
  - issues significant to accident frequencies, containment performance, and risks,
  - risk-significant uncertainties which may merit further research,
  - comparison with NRC's safety goals (cf. Sect. 6.6); and
  - the potential benefits of a severe accident management programme in reducing accident frequencies;
- to provide a set of PRA models and results which can support the ongoing prioritization of potential safety issues and related research.

The five commercial nuclear power stations investigated are:

1. Unit 1 of the Surry Power Station (already analysed in the RSS).
2. Unit 1 of the Zion Nuclear Power Plant.
3. Unit 1 of the Sequoyah Nuclear Power Plant.
4. Unit 1 of the Peach Bottom Atomic Power Station (already analysed in the RSS).
5. Unit 1 of the Grand Gulf Nuclear Station.

The risk analyses cover the following topics:

- expected frequencies of severe accidents;
- performance of the containment and other mitigative systems and structures in such accidents;
- offsite consequences (health effects, property damage etc.) of such accidents.

Both internal and external initiating events were considered in the analyses of the Surry 1 and Peach Bottom 2 power plants. In the case of the remaining power stations only internal initiating events were studied. The core-damage frequencies are characterized by subjective probability density functions, as shown in Table 3.5.

The comparison of the core-damage frequencies calculated in [1] for the Surry and Peach Bottom plants with those from the RSS shows that lower medians and wider 90% confidence intervals result in the new study. The lower median values reflect the fact that hardware modifications and improved procedures were implemented in both plants since the RSS was completed. The increased level of detail in modelling of the new study introduced more uncertainties than it reduced or eliminated. This explains the wider confidence intervals.

Table 3.6 shows the contributions of the individual accident sequences to the mean core-damage frequency for each of the plants.

The accident progression and containment performance analyses determine the modes and frequencies of containment failures under the loads from core-melt accidents. Knowledge necessary to calculate all possible courses of acci-

Table 3.5. Annual core-damage frequencies for internal events from [1] and comparison to RSS

Plant	5%	median	mean	95%
Surry				
[1]	$6.8 \cdot 10^{-6}$	$2.3 \cdot 10^{-5}$	$4.1 \cdot 10^{-5}$	$1.3 \cdot 10^{-4}$
RSS	$2.9 \cdot 10^{-5}$	$4.6 \cdot 10^{-5}$	–	$3.0 \cdot 10^{-4}$
Peach Bottom				
[1]	$3.5 \cdot 10^{-7}$	$1.9 \cdot 10^{-6}$	$4.5 \cdot 10^{-6}$	$1.3 \cdot 10^{-5}$
RSS	$1.8 \cdot 10^{-5}$	$2.6 \cdot 10^{-5}$	–	$1.1 \cdot 10^{-4}$
Sequoyah	$1.2 \cdot 10^{-5}$	$3.7 \cdot 10^{-5}$	$5.7 \cdot 10^{-5}$	$1.8 \cdot 10^{-4}$
Grand Gulf	$1.7 \cdot 10^{-7}$	$1.2 \cdot 10^{-6}$	$4.0 \cdot 10^{-6}$	$1.2 \cdot 10^{-5}$
Zion	$1.1 \cdot 10^{-4}$	$2.4 \cdot 10^{-4}$	$3.4 \cdot 10^{-4}$	$8.4 \cdot 10^{-4}$

Table 3.6. Contributions to core-damage frequency (mean values) (from [1])

Type of event	Plant Surry	Peach Bottom	Sequoyah	Grand Gulf	Zion
Internal events	$4.1 \cdot 10^{-5}$	$4.5 \cdot 10^{-6}$	$5.7 \cdot 10^{-5}$	$4.0 \cdot 10^{-6}$	$3.4 \cdot 10^{-4}$
Station blackout	$2.2 \cdot 10^{-6}$	–	$3.9 \cdot 10^{-6}$	–	–
Short term	$5.4 \cdot 10^{-6}$	–	$9.6 \cdot 10^{-6}$	–	–
Long term	$2.2 \cdot 10^{-5}$	–	$5.0 \cdot 10^{-6}$	–	–
ATWS	$1.6 \cdot 10^{-6}$	$1.9 \cdot 10^{-6}$	$1.9 \cdot 10^{-6}$	$1.1 \cdot 10^{-7}$	–
Transient	$2.1 \cdot 10^{-6}$	$1.4 \cdot 10^{-7}$	$2.5 \cdot 10^{-6}$	–	–
LOCA	$6.0 \cdot 10^{-6}$	$2.6 \cdot 10^{-7}$	$3.6 \cdot 10^{-5}$	–	–
Interfacing-systems LOCA	$1.6 \cdot 10^{-6}$	–	$6.5 \cdot 10^{-7}$	–	–
STGR	$1.8 \cdot 10^{-6}$	–	$1.7 \cdot 10^{-6}$	–	–
External events					
Seismic (LLNL)	$1.2 \cdot 10^{-4}$	$7.7 \cdot 10^{-5}$	–	–	–
Seismic (EPRI)	$2.5 \cdot 10^{-5}$	$3.1 \cdot 10^{-6}$	–	–	–
Fire	$1.1 \cdot 10^{-5}$	$2.0 \cdot 10^{-5}$	–	–	–

ATWS: Anticipated transients without scram.
LOCA: Loss-of-coolant accident.
Interfacing-systems LOCA: LOCA in a connecting pipe to the primary circuit.
STGR: Steam generator tube rupture.
LLNL: Calculations performed by the Lawrence Livermore National Laboratory.
EPRI: Calculations performed by the Electric Power Research Institute.

dent progression on the basis of validated mechanistic computer programs is currently not available. To bridge gaps in knowledge expert judgement was used to supplement and interpret information and data on unresolved issues. Figure 3.6 shows the frequencies of early containment failure and bypass together with the corresponding 90% confidence intervals. These are the failure modes which lead to the most severe offsite consequences. Table 3.7 gives the probabilities for different containment failure modes under the condition that core-damage

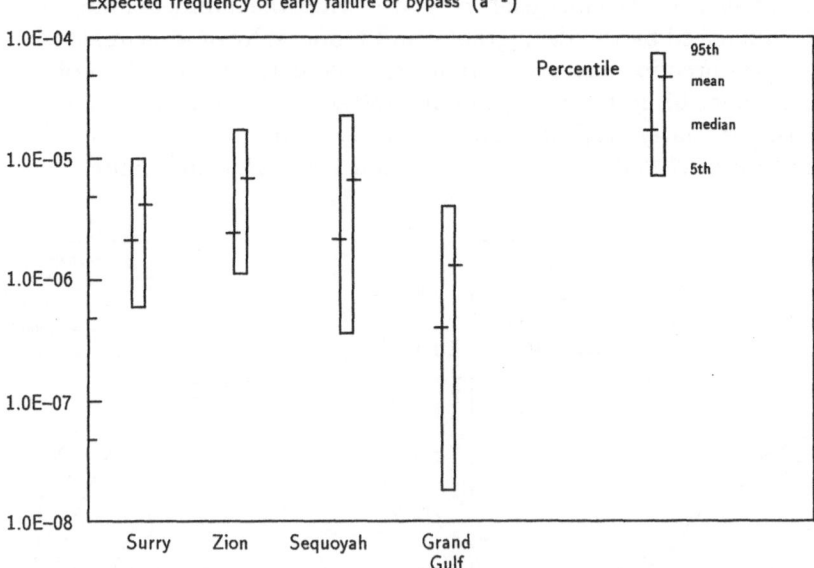

Fig. 3.6. Frequencies of accident sequences with early containment failure or containment bypass at all plants (from [1])

Table 3.7. Mean conditional probability of containment failure for three PWRs (from [1])

Containment failure mode	Surry	Zion	Sequoyah
Early failure with reactor vessel at pressure > 13.8 bar	0.004	0.02	0.04
Early failure with reactor vessel at pressure < 13.8 bar	0.0	–	0.02
Late containment failure	< 0.01	–	0.04
Containment bypass	0.12	0.006	0.06
Others (steam explosion, basemat melt-through)	0.06	0.22	0.18
Vessel breach, no containment failure	0.34	0.76	0.26
Accident progression arrested before vessel breach	0.46	0.76	0.38

has occurred. It is important to note that the probability of the containment not failing is significant, and that the probability of early containment failure is shown to be low.

Previous remarks on the knowledge base also hold true with regard to the determination of the release of radioactive substances to the environment, the source term calculations. There are large uncertainties associated with the results; the 90% confidence intervals range over several orders of magnitude.

Offsite consequences are largely determined by the core-damage frequencies, the radionuclide release parameters of the source term, and the meteorological parameters applicable to the site in question. The number of early fatalities additionally depends on the population distribution in the vicinity of the plant and the effectiveness of the emergency response, while the number of late cancer fatalities is influenced by the demography and topography of the site, the land use, agricultural practice and productivity as well as the distribution of fresh water at a distance of up to 50 or 100 miles (80–160 km) from the plants.

The CCDFs for early fatalities and late cancer fatalities as measures of risk are shown in Figs. 3.7 and 3.8 for the Surry and Peach Bottom plants.

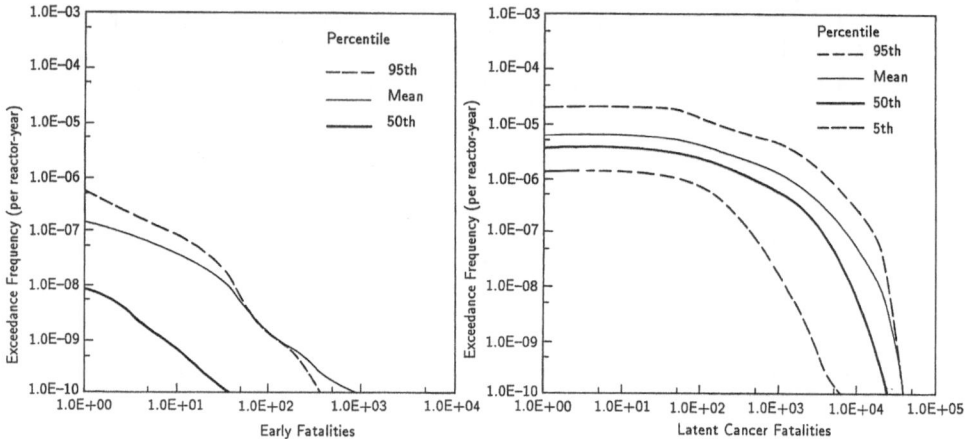

Fig. 3.7. Risks from early and latent cancer fatalities at Surry – internal events (from [1])

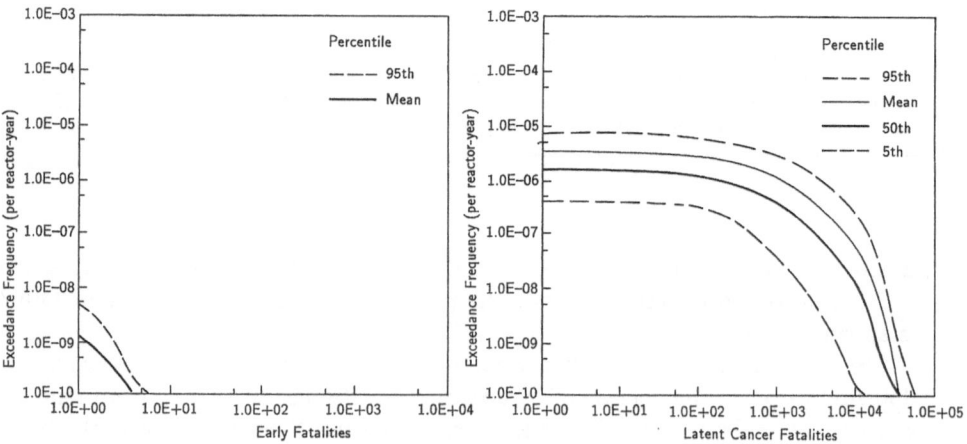

Fig. 3.8. Risks from early and latent cancer fatalities at Peach Bottom – internal events (from [1])

3.2.4 Risk Study for German Nuclear Power Plants

3.2.4.1 Phase A

In the German Risk Study, Phase A (DRS-A [16]) the assumptions and methods of the RSS were used for assessing the risks from severe accidents in nuclear power stations in the FRG. It was completed in 1979. Direct application of the results of the RSS to conditions in the FRG was not considered adequate, because

- there are significant differences between the reference plants of the US study and the German plants. These differences chiefly concern the design and the functioning of the safety systems,
- the population density in the FRG and in central Europe is higher than in the US (average population density: 248 inhabitants per km^2 in the FRG, as opposed to 25 inhabitants per km^2 in the US).

The reference plant of the German risk study is the Nuclear Power Station Biblis B [27], which is equipped with a four-loop pressurized water reactor of 1300 MWe designed by KWU. The analysis comprises all the steps described in Chap. 2.

Consequences of the potential releases of radioactive materials into the environment from the power plant were estimated, together with their expected frequencies of occurrence. All sites with commercial light water reactors exceeding a power of 600 MWe which were in operation, under construction, or subject to licensing in July 1977 were considered. This led to 19 sites with 25 power stations, all of which were supposed to be of the type of the reference plant.

In selecting the initiating events the DRS-A followed the RSS. The expected frequencies of core-melt found in both studies are given in Table 3.8. Figure 3.9 shows the relative contributions of the different initiating events to the core-melt frequencies.

Table 3.8. Expected core-melt frequencies per reactor year

	DRS-A [16]	RSS [15] (PWR)
Median	$4 \cdot 10^{-5}$	$5 \cdot 10^{-5}$
Mean	$9 \cdot 10^{-5}$	$1 \cdot 10^{-4}$

The most important contribution to the core-melt frequency stems from small leaks in the primary coolant system. Transients together with a failure in the open position of the relief valves of the pressurizer also contribute substantially. This accident sequence is similar to the accident at TMI and had

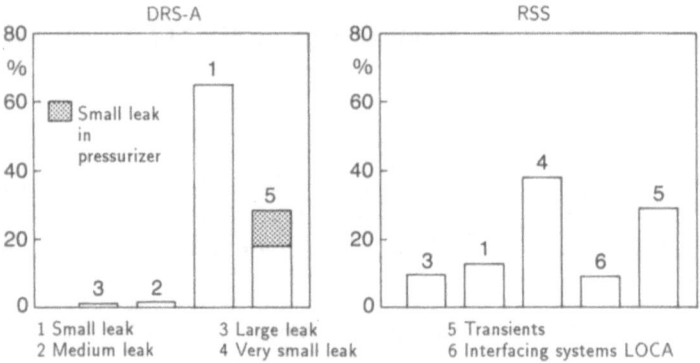

Fig. 3.9. Relative contributions of different initiating events to the core-melt frequencies

already been discussed in an early phase of the analysis [28]. The identification of this event sequence led to several improvements in the system which had been implemented in German plants one year before the accident at TMI occurred.

No detailed analysis of the performance of the containment was carried out. In-vessel steam explosions and failures to isolate containment after core-melt resulted in the severest impact on the environment. In the analysis a conditional probability of 0.01 was used for the occurrence of a powerful steam explosion. If the containment is not destroyed by a steam explosion and if it is properly isolated, its failure was calculated to occur 25 hours after the onset of core-melt as a consequence of overpressure as opposed to 10 hours in the RSS. The longer time to failure for the German plant results from the large dry containment, which fails at approximately 8 bar.

Accident consequences were assessed for the already mentioned 19 sites in the FRG with 25 nuclear power plants. The calculations were based on a circular area with a diameter of 2500 km, inhabited by 670 000 000 people.

Early fatalities can occur up to a distance of 20 km from the sites. However, based on the assumptions concerning containment loads and its retention capability, in 99% of all cases no early fatalities are to be expected. A maximum number of 16 600 early fatalities occurring with an expected annual frequency of $4.8 \cdot 10^{-10}$ were estimated.

As a consequence of the linear dose-effect relationship whithout a threshold that was used in the DRS-A, late fatalities are expected to occur wherever the dispersion of the released substances leads to radioactive exposure. Therefore a cancer risk is calculated even for those doses which are below the 0.05 Sv (5 rem) limit of the Radiation Protection Ordinance of the FRG (cf. Sect. 6.3.2). A maximum number of 107 800 late fatalities was estimated to occur with an expected annual frequency of $4.8 \cdot 10^{-10}$.

The risks of early and late fatalities are presented in terms of CCDFs. Figures 3.10 and 3.11 provide a comparison of the results of the RSS with those of the DRS-A. In both cases the results were normalized to 25 power plants.

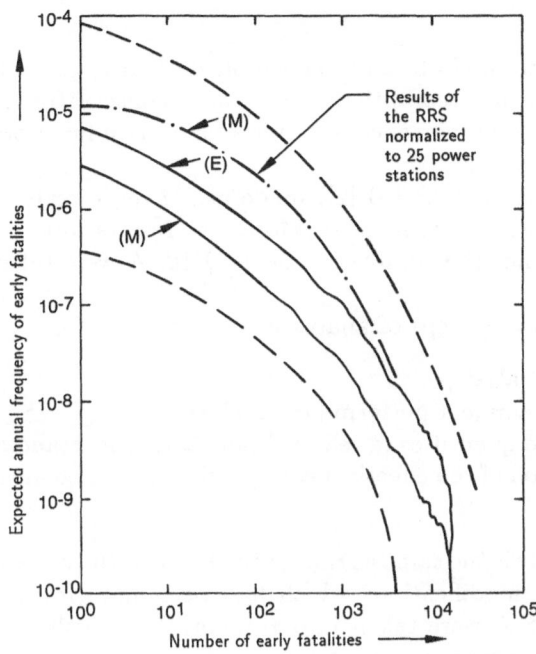

Fig. 3.10. CCDFs for early fatalities from RSS and DRS-A normalized to 25 plants. Curve E: reference curve from the DRS-A (expected values). Curve M: Medians from the DRS-A in analogy with the RSS (from [29])

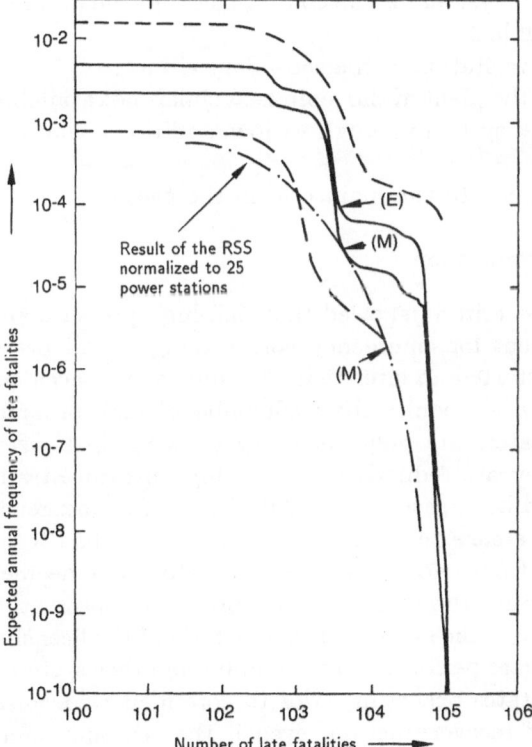

Fig. 3.11. CCDFs for late fatalities (collective risk) from the RSS and the DRS-A normalized to 25 plants. Curve E: reference curve from the DRS-A (expected values). Curve M: Medians from the DRS-A in analogy with the RSS (from [29])

3.2.4.2   Phase B

In the DRS-A the methods used in the RSS were to be applied as far as possible which explains why the subsequent assumptions made were conservative in many respects. In particular, the minimum success criteria for system functions from the licensing procedure were used.

The German Risk Study, Phase B (DRS-B [2]) made use of more realistic assumptions for analysing the accident sequences. Moreover, results from reactor safety research obtained since the completion of the DRS-A were taken into account.

The study comprises the following steps of analysis:

- event sequence and systems analysis,
- accident progression and containment performance analysis,
- determination of the types and quantities of released radioactive substances,
- identification and quantification of the effectiveness of accident-management measures.

The reference plant of the study is the same as that of the DRS-A, the nuclear power station Biblis B. System modifications which had been implemented since the completion of the DRS-A were taken into account. The results are significantly influenced by the following modifications:

- installation of a semi-automatic system for the controlled cooldown at a rate of 100 K/h in the case of small leaks,
- improvements of the relief valve station of the secondary circuit,
- automatic partial cooldown of the plant, if the main heat sink is not available,
- control of the pressurizer relief system through various additional isolation signals,
- possibility of restoring connection to the main grid in the case of a failure of the emergency diesels,
- installation of a backup grid connection.

Analysis of the system success criteria showed that one high pressure and one low pressure train is sufficient for emergency core cooling for all break sizes, if the reactor cooldown is started no later than 30 minutes after accident initiation. Not all of the break sizes require the availability of high pressure injection systems. The accumulators are only necessary for some leak sizes; in the 200 to 500 cm^2 range they are redundant for the high pressure train. In contrast, in the DRS-A it had been assumed that for successful emergency core cooling at least two high pressure and two low pressure trains had to be available. In addition, injection from two or more accumulators was deemed necessary. These examples illustrate the elimination of previous conservative assessments resulting from the use of the system success criteria of the licensing procedure. Similar calculations were performed for a number of other systems.

In the study, plant damage states occurring prior to core-melt were analysed. Most of them still permit recovery action, even if the relevant safety systems should fail. The accident will only progress to core-melt if recovery is not successful. In the DRS-A, on the other hand, all core-damage states were considered as core-melt.

Table 3.9 presents the initiating events analysed in the study and Fig. 3.12 shows the contributions of principal accident sequences to the frequency of plant damage. It also illustrates how the unavailability of certain systems contributes to the result. Fig. 3.13 shows the distribution function of the overall frequencies of plant damage states from internal initiating events and indicates some characteristic values used to describe the result of a probabilistic analysis, namely 5% and 95% percentiles, mean, and median.

Table 3.9. List of initiating events analysed in the DRS-B and their expected frequencies of occurrence

Designation	Annual expected frequency of occurrence
Leaks from the primary circuit	$3.2 \cdot 10^{-3}$
Leaks at the pressurizer due to transients	$1.8 \cdot 10^{-4}$
Small leak at the pressurizer ($40\,\mathrm{cm}^2$) after erroneous opening of the safety valve	$8.5 \cdot 10^{-4}$
Leak from the primary circuit in the annulus (2–$500\,\mathrm{cm}^2$)	$< 10^{-7}$
Steam generator tube rupture	$6.5 \cdot 10^{-3}$
Operational transients	0.93
Transients due to leaks from the live steam circuit	$7.8 \cdot 10^{-4}$
Operational transients with failure of scram	$3.9 \cdot 10^{-5}$
Global plant internal events	$8.9 \cdot 10^{-6}$
External events (earthquake and aircraft crash)	$8.0 \cdot 10^{-4}$

The overall frequency of plant damage states from internal events is $2.6 \cdot 10^{-5}\,\mathrm{a}^{-1}$*. It is lower than the frequency estimated for core-melt in the DRS-A ($9 \cdot 10^{-5}\,\mathrm{a}^{-1}$), although several additional initiating events were taken into account. This reflects the effects of the improvements to the systems implemented since the completion of the DRS-A and of the use of more realistic success criteria.

Accidents initiated by fires, earthquakes, aircraft crashes, and the flooding of the annulus between the steel containment and the steel-concrete reactor building were analysed as well. Their contribution to the frequency of plant damage states is smaller than $3.6 \cdot 10^{-6}\,\mathrm{a}^{-1}$.

As previously mentioned, accident progression from plant damage states to core-melt is only possible if all active recovery actions fail. But even then accident progression could still be arrested by passive physical phenomena before the bottom head of the pressure vessel fails. However, this possibility was not explored. Active accident management procedures were accounted for in the

---

* a^{-1} or $1/\mathrm{a}$ means per year.

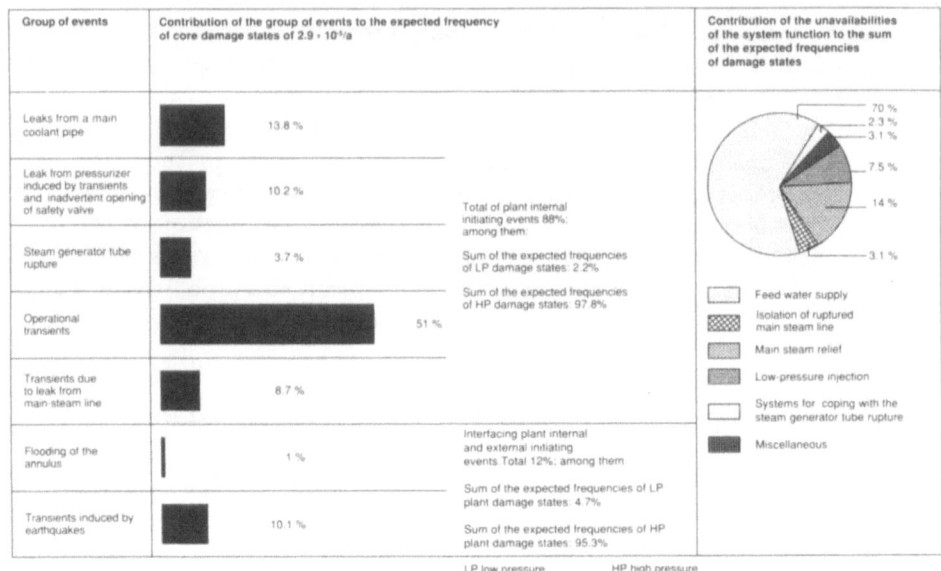

Fig. 3.12. Contributions of the different accident sequences to the overall frequency of plant damage states

study, in particular bleed and feed measures to be carried out both in the primary and secondary circuits. Only a preliminary quantification was performed assuming that in 99 out of 100 cases core cooling could be restored, unless unfavourable boundary conditions pointed to a lower success rate. With these measures included, 88% of the plant damage states can be transformed into a safe state, with 12% resulting in core-melt. Thus the annual expected frequency of core-melt is reduced to $3.6 \cdot 10^{-6}$. In addition, the frequency of the sequences of high pressure core-melt (cf. Sect. 3.1.1.3) is reduced. Only 12% of all core-melt cases take place at high system pressure ($4.5 \cdot 10^{-7} \, a^{-1}$).

Table 3.10 shows the time elapsing from the onset of the accident to core-melt and pressure vessel failure. Figures 3.14 and 3.15 present the relative contributions of the principal accident sequences to core-melt frequencies at low and high system pressures.

In the analyses of accident progression and containment performance the effects of high pressure ejection of molten core material from the failing pressure vessel, of deflagrations and detonations of hydrogen and of steam explosions were studied on the basis of new research results. Each of these phenomena may cause early containment failure and hence severe offsite consequences. In a current research programme the feasability of hydrogen control by igniters and surfaces coated with catalytic foils are investigated. Steam explosions with a potential for threatening the containment are considered to contribute insignificantly to risk.

The following containment failure modes were investigated in the study:

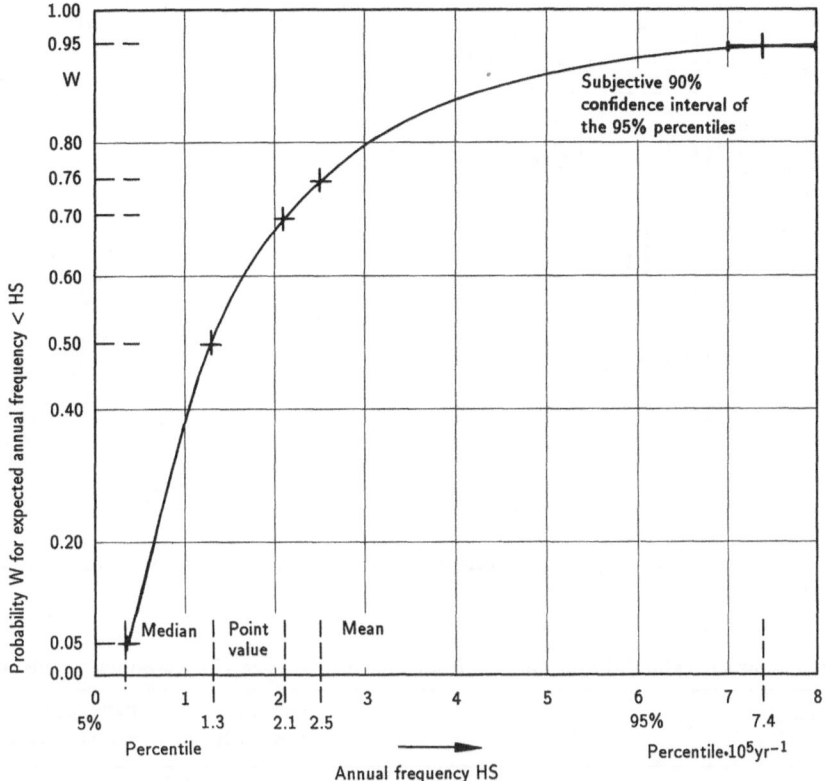

Fig. 3.13. Distribution function of the plant damage state frequency from internal initiating events

- large leak from the containment above basemat level, e.g. due to high system pressure core-melt, H_2 deflagration or detonation, aircraft crash;
- containment bypass due to a leak from the primary system in the containment annulus;
- containment bypass due to a leak in the steam generator at low pressure in the primary circuit;
- leak from the containment (area: $10\,\mathrm{cm}^2$) from the beginning of the accident sequence;
- controlled containment venting;
- containment basemat failure.

In view of the many uncertainties associated with the analysis of containment performance, the conditional probabilities of the different containment failure modes were not calculated. Only bounds of the frequencies of occurrence of the six different failure modes of the containment were estimated. The magnitude of the fission product release associated with the containment failure modes is shown in Table 3.11.

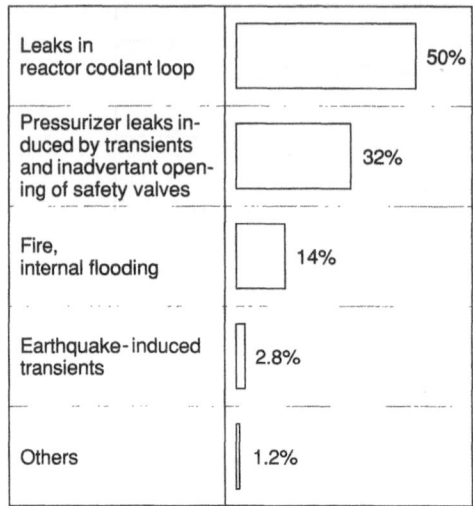

Leaks in reactor coolant loop	50%
Pressurizer leaks induced by transients and inadvertant opening of safety valves	32%
Fire, internal flooding	14%
Earthquake-induced transients	2.8%
Others	1.2%

Fig. 3.14 Contributions of the principal accident sequences to the frequency of core-melt at low pressure

Table 3.10. Characteristic times in minutes from accident initiation for core-melt accidents

Core-melt case	Onset of core-melt	Failure of pressure vessel
Low pressure in primary circuit	55	120
High pressure in primary circuit	110	140
Low pressure in primary circuit after depressurization by accident management measures (ND*)	330	410
Steam generator tube rupture and ND*	540	710
Leak from primary circuit in the annulus, low pressure in primary circuit	80	140

Table 3.11. Fission product releases normalized to the core inventory

Accident sequence	Time (h)	Kr-Xe	I	Cs	Te	Sr
1 Early failure of containment	> 2	1.0	> 0.5	> 0.5	> 0.5	0.4
2 Interfacing systems loss-of-coolant accident	> 1.5	1.0	0.37	0.37	0.23	0.17
3 SGTR[a]	8	0.17	$2.5 \cdot 10^{-2}$	$2.5 \cdot 10^{-2}$	$1.5 \cdot 10^{-2}$	$1 \cdot 10^{-5}$
4 Small leak in containment	6	1.0	$8 \cdot 10^{-3}$	$4 \cdot 10^{-4}$	$2 \cdot 10^{-3}$	$2 \cdot 10^{-4}$
5 Filtered venting[b] of containment	> 4 d	0.9	$2 \cdot 10^{-3}$	$3 \cdot 10^{-7}$	$4 \cdot 10^{-6}$	$2 \cdot 10^{-7}$

[a] with water pool on secondary side of defect steam generator (in-vessel phase)
[b] aerosol filter

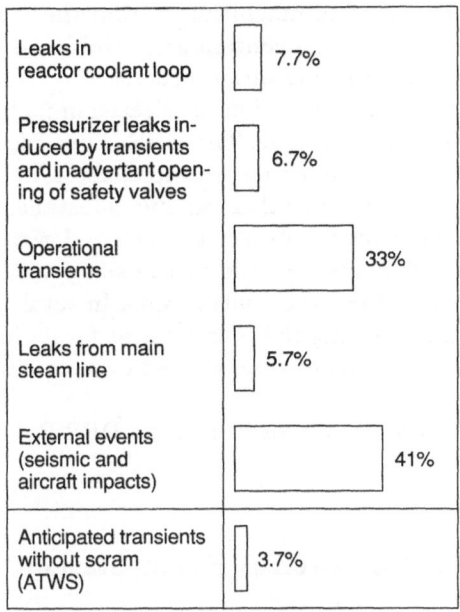

Fig. 3.15. Contributions of the principal accident sequences to the frequency of core-melt at high pressure

Late overpressure failure of the containment is expected after four to five days, in contrast to one day calculated in the DRS-A. In particular, this results from a better understanding of the core-melt-concrete interaction. Accordingly, ample time is available for the preparation of accident-management measures which can delay or even prevent overpressure failure of the containment. One of the possibilities considered is controlled venting, which can be performed by using available or readily installable equipment. Controlled venting retains radioactive substances inside the containment because of the depletion of the containment atmosphere due to deposition processes (cf. Sect. 3.1.1.4). By using filters a further reduction of the radioactivity released to the environment is possible.

Investigations into the behaviour of iodine and aerosols after the release into the containment led to improved theoretical models. After a core-melt accident the containment atmosphere consists of a mixture of steam and gases, suspending aerosol particles of up to 20 g mass concentration per m^3. For example, in the case of a low pressure core-melt accident, there may be between 1 and 3 tonnes of dispersed airborne mass in the containment at the onset of the core-melt-concrete interaction. More than 95% are non-radioactive; however, the major portion of radioactive substances is bound to aerosol particles. The initial aerosol concentration is reduced through deposition and sorption processes and is decreased by five to six orders of magnitude within the four or five days preceding late overpressure failure of the containment. Noble gases and the gaseous iodine do not participate in this deposition process.

In the DRS-A it had been assumed that nearly all iodine is released in its biologically most noxious form, namely elementary iodine. The DRS-B considers the most important physical and chemical reactions; for example, the formation of organic iodine compounds with paints and other organic substances, the exchange processes of gaseous iodine species between the gas and water phases, and the formation of water-insoluble silver iodide in the sump water through a reaction of I_2 with metallic silver from the control rods. In total, it can be stated, that in the case of controlled venting the source terms for iodine and cesium are about one order of magnitude lower than for the overpressure failure considered in the DRS-A.

Calculations for offsite consequences were not performed in the DRS-B.

3.2.5 Probabilistic Analysis for the Sizewell B Power Station

The proposed power plant Sizewell B (United Kingdom) is to be equipped with a four-loop PWR of 1245 MWe power of Westinghouse design. The probabilistic study for this power station was carried out by Westinghouse and several consulting companies [30]. It comprises all the steps from identification of the initiating events to the determination of release categories and the quantification of their expected frequencies of occurrence (cf. Chap. 2). Accident-consequence calculations for the plant were performed by the National Radiological Protection Board [31]. The boundary conditions for these calculations were taken from [30]. Some of the assumptions had been corrected by the Central Electricity Generating Board (CEGB), because they were considered to be too conservative. Further corrections and additional calculations were made for the Sizewell Public Inquiry [32].

In both studies, realistic assumptions were employed in the calculations. In general, point values instead of probability distributions were used for uncertain parameters. Sensitivity analyses were performed for the accident-consequence calculations [33]; dependence of the results on variations of the input parameters turned out to be insignificant in most cases [32]. External events were not considered in the study.

The fault trees were quantified with data from general operating experience. Plant-specific data were generated using Bayesian analyses [34]. Common cause failures were treated in accordance with the design safety criteria of the CEGB (cf. Sect. 6.6.4), which require the use of a lower bound for the probabilities of common cause failures of 10^{-5}, and of 10^{-3} for total unavailability of the individual systems, even if the calculations yield lower values.

In general, a value of 10^{-4} per demand was used as common cause failure probability; only in particular cases was the lower bound of 10^{-5} taken.

Errors in operator actions were modelled with human reliability event trees; they were quantified using the THERP method (cf. Sect. 2.1.4.8 and [35]).

10^{-4} was used as the conditional probability for the occurrence of a steam explosion in the core-melt sequence. In [32] no reasons were seen, particularly in view of the results in [36], to narrow the range $[10^{-1}, 10^{-4}]$ used in the RSS. Sensitivity studies using a probability of 10^{-2} instead of 10^{-4} resulted in an increase of the number of early deaths by 13% and of that for fatal cancers by 21% [32].

A failure of the containment as a result of a hydrogen explosion is considered unlikely in [30]. In [33], however, doubts concerning this are expressed.

The broad spectrum of potential releases resulting from core-melt accidents was discretized into 12 release categories. The fractions of the core inventory released from the containment were determined using the methods of the RSS. Since these results are generally considered to be too conservative, more realistic calculations accounting for new research results were carried out for several dominant release categories. They show a reduction of the releases in comparison to the results obtained with the previous methods.

Dispersion of radioactive substances was treated by the Gaussian model using site-specific meteorological parameters. Evacuation measures were accounted for. Early deaths, prodromal vomiting, fatal cancers, hereditary deaths, non-fatal thyroid cancers, rejection of milk, and the number of persons to be evacuated are the categories of damage whose magnitudes and expected frequencies of occurrence were assessed.

Frequency of core-melt due to internal events was estimated to be $1.24 \cdot 10^{-6}$ per reactor year [32] ($1.16 \cdot 10^{-6}$ in [30]). This value lies within the uncertainty range of the result from the preconstruction safety report [37] where a core-melt frequency of $3.1 \cdot 10^{-6}$ per reactor year had been assessed. It should be mentioned that the two studies were performed by different teams; their scope is also different.

Significant contributions to the frequency of core degradation stem from the loss of coolant through a small leak (35.4%), from large and medium size loss-of-coolant accidents (35.6%), and from component failures after an anticipated transient without scram (ATWS) (11.0%).

With 94% of the accident sequences the containment is expected to remain as a barrier against the release of radioactive substances. The total frequency of releases with containment failure or bypass was calculated to be $7.5 \cdot 10^{-8}$ per year. 61.1% of the releases are due to early containment failures or containment bypass, 13.5% result from late overpressure failure of the containment. Only 1.8% of the early failures of the containment are due to steam explosions. 23.7% are caused by a melt-through of the basemat of the containment. A survey of additional results of the studies is given in Figs. 3.16, 3.17, and 3.18.

In [33] and [30] the conclusion is made that the study demonstrates compliance with the essential design safety criteria of the CEGB (cf. Sect. 6.6).

Some of the uncertainties of the results were addressed by means of sensitivity studies in [32], which led to corrections of the original figures. Applied to the

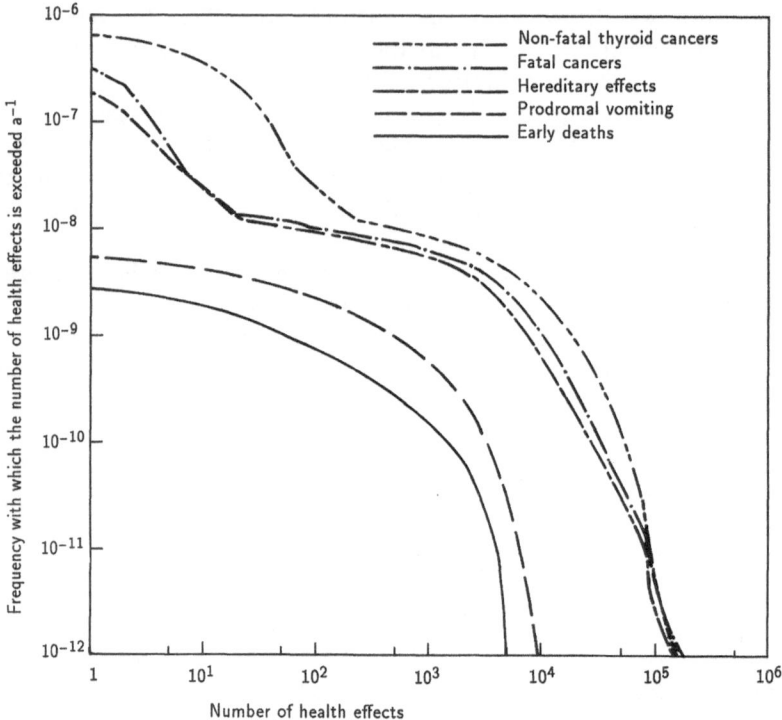

Fig. 3.16. Frequency distributions of the incidence of each health effect from degraded core accidents per year of reactor operation (from [32])

probabilities for hydrogen burning, this produced an increase of the expected number of early deaths of 4% and of that for fatal cancers of 6%. Analyses on containment strength and corresponding sensitivity studies indicate an increase of the expected number of early deaths by 7% and of that of fatal cancers by 9%. Using revised frequencies for initiating events and reliability data the expected number of early deaths is reduced by 8% and that of fatal cancers by 12%.

In summary, corrections of the initial investigation [30] and the results of the sensitivity studies did not have a major impact on the global results, although the expected frequencies of occurrence of individual release categories changed.

3.2.6   Analyses of Precursors of Severe Core-Damage Accidents

The study on precursors of severe core-damage accidents [38] was prepared as a result of the recommendation of the Lewis Review Group [22] to make more use of operating experience and of event sequence and fault tree analyses for assessing the risks of nuclear power plants.

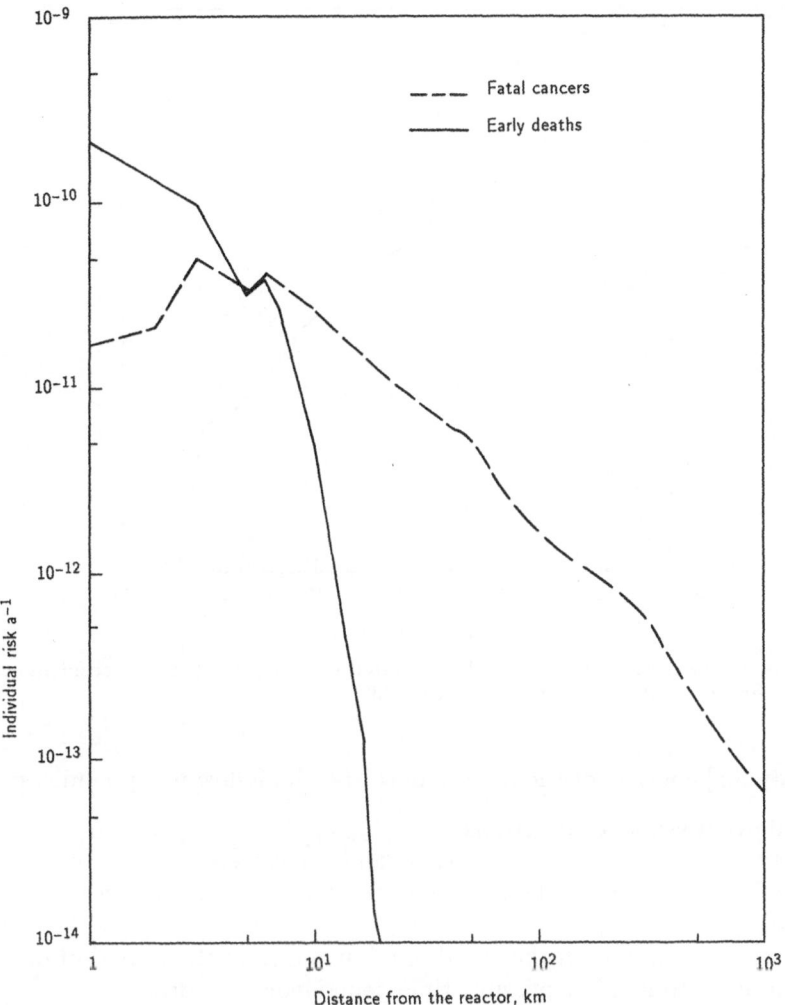

Fig 3.17. Individual risk from degraded core accidents per year of reactor operation (from [32])

Data bases for the precursor study are licensee event reports. Events which could be viewed as potential precursors to severe core-damage were evaluated on the basis of statistical data from the years 1969 to 1979. A more recent report was prepared using data from the years 1980 to 1981 [39]. Refining the approach of the first report, plant-specific properties and the possibility of carrying out suitable accident management measures in the plants were considered. In addition, technical differences between the different plants were taken into account. Uncertainties of the results are indicated. Compared to the first study, somewhat lower frequencies for the occurrence of core-melt were estimated, namely a 95% confidence limit of $5.8 \cdot 10^{-4} \, \text{a}^{-1}$.

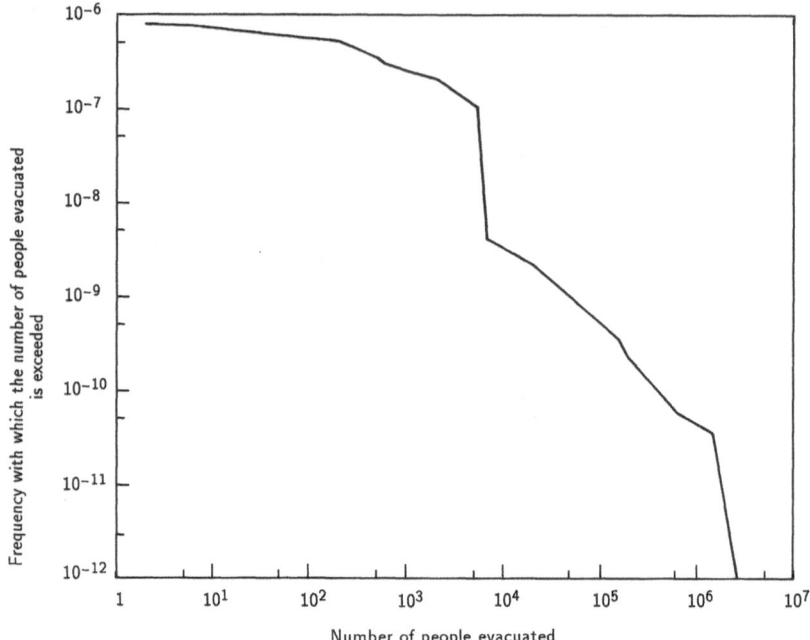

Fig. 3.18. Frequency distribution of people evacuated due to degraded core accidents per reactor year – first estimate release fractions (from [32])

The study [38] consists of the main report and the following appendices:

A: standard event sequence diagrams.
B: compilation of data for precursor events and event trees.
C: additional information on the selection of potential precursors regarded as important.
D: supporting information on the trend of failures during the period of observation in order to identify possible time dependencies of data.
E: additional information on the trend analyses.

On the whole 19 400 licensee event reports were reviewed and 500 of them were selected for more detailed examination, of which 196 remained as potential precursors. These were used to estimate the failure or success probabilities in the event trees[2].

This led to expected core-melt frequencies between $1.7 \cdot 10^{-3}$ and $4.5 \cdot 10^{-3}$ per reactor year.[3] A time dependence of the occurrence of significant events or variations between plants of different vendors could not be inferred from the available data. The accident at TMI contributes about 50% to this esti-

---

[2]   These probabilities are calculated in risk studies with fault tree analyses on the basis of reliability data for technical components and human actions (cf. Chap. 2).

[3]   These are not confidence bounds but the results of different interpretations of the underlying data.

mate. Another 32% stems from two further events, the cable fire at the Browns Ferry power station and the failure of the non-nuclear instrumentation and subsequent dry out of the steam generators at the Rancho Seco power plant. Without these three events a core-melt frequency of $7.7 \cdot 10^{-4}$ per reactor year is expected. This illustrates the extent to which the results of the study are influenced by the dearth of operating experience (432 years of reactor operation form the basis of the study). The selection criteria used for the precursors are the same as if instead of core-melt severe damage to the fuel rod cladding or to the core had been investigated.

The authors emphasize the large uncertainties of their results. However, the results of the study are not inconsistent with theoretically estimated core-melt frequencies, as is shown in Fig. 3.19.

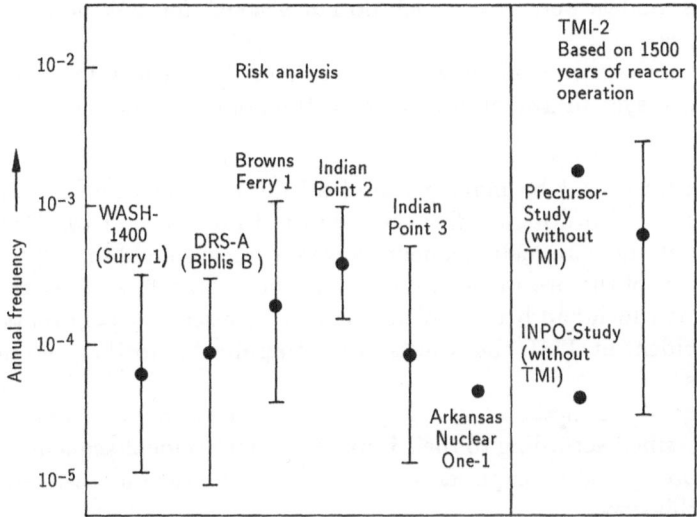

Fig. 3.19. Results for the core-melt frequencies from several risk analyses with the corresponding uncertainties

A time dependence of the occurrence of significant events or variations between plants of different vendors could not be inferred from the available data.

The methodology and results of the precursor study [38] were reviewed by the Institute of Nuclear Power Operation [40]. The following topics were chiefly criticized:

- often no credit was given for plant systems which were available for accident control, even if they had been used in some cases to terminate event sequences which could have led to core-melt;
- in several cases reliability of the systems was underestimated, because of incomplete information or erroneous assumptions about the failure modes investigated;
- the ability of the operators to repair failed systems quickly was underestimated;

- according to the investigations in [38] the accident at TMI, the cable fire at Browns Ferry, and the failure of the non-nuclear instrumentation at Rancho Seco contribute 85% to the total core-melt frequency. In [40] core-melt frequencies were estimated which are a hundred times lower in the case of Browns Ferry, and a thousand times lower for Rancho Seco. Significant improvements to the plants which were made after the events were not taken into account;
- the calculations in [38] are based on 432 years of reactor operation, although accumulated operating experience at the time of publication of the report amounted to 1500 years. During this period of time one accident with partial core-melt (TMI) had occurred;
- the evaluation of the events is based on the licensee event reports; their condensed representation led to misinterpretations in some cases;
- in [38] simplified models were used which do not account for specific properties of the plants;
- simplifying assumptions were made about operational data, which led to an underestimation of system availabilities and of the ability of the operators to intervene.

In [40] it is concluded that by making use of all the available information core-melt frequencies which do not differ significantly from those of the RSS would result and that the core-melt frequencies were overestimated in [38] for the time of occurrence of the precursors. Since the authors did not consider the system modifications which had been implemented as a consequence of events, particularly the accident at TMI, the results are not applicable to the present status of the plants.

It is an advantage of the method that its systematic approach allows abnormal events to be classified according to their importance for accident sequences. In the long run more operating experience will enable analytical assessments to be validated in this way.

In the FRG, the Gesellschaft für Reaktorsicherheit (GRS) carried out analyses of precursors of accidents which would potentially cause severe core-damage [41]. In view of the critique of the US precursor study and of the limited operating experience available in German nuclear power plants, its methodology to determine the mean values of the frequencies of occurrence of initiating events and the mean values of probabilities of system failures was not adopted in the German precursor study. The study was performed for the nuclear power plant Biblis (Units A and B) [27]. The approach differs from the US precursor study in that

- plant-specific frequencies of the occurrence of initiating events were estimated on the basis of more detailed data than the licensee event reports,
- the system unavailabilities were obtained from analytically assessed plant-specific unavailabilities of redundant system trains as well as from observed system failures, potential system failures or multiple system failures, which had been observed in Biblis,

– human actions were quantified on the basis of plant-specific operating experience.

The minimum success criteria for 100% power from the licensing procedure were used, just as in the DRS-A. Figure 3.20 shows the observed frequencies of precursors to severe core-damage. The time history of the estimated expected frequencies of severe core-damage, as determined from the precursors, is shown in Fig. 3.21. The significant decrease of the frequencies as the time of operation increases can be attributed to numerous system improvements and to the growing experience of the operating personnel. The mean value, averaged over the total period of operation, is estimated to be $5 \cdot 10^{-5}$ a^{-1}. According to the results of an uncertainty analysis, the expected frequency of severe core-damage,

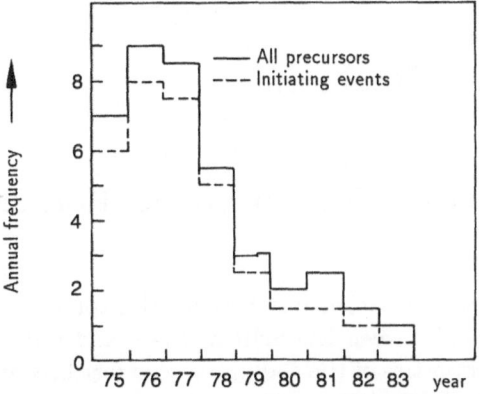

Fig. 3.20. Frequency of precursors per reactor year over several years of operation

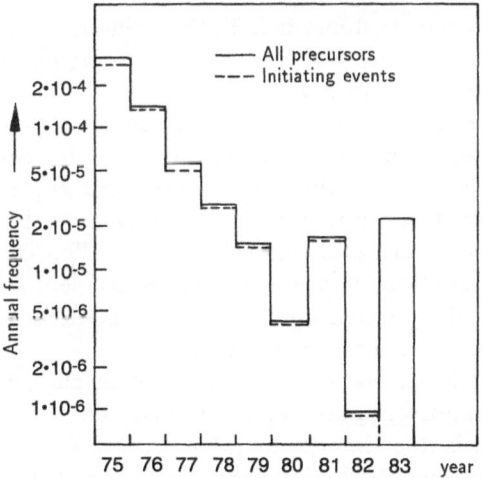

Fig. 3.21. Annual expected frequency for severe core-damage assessed from precursors over several years[4]

---

[4] In the strict sense this is the expected frequency of non-compliance with the success criteria of the licensing procedure.

based on observed precursors, is below $5 \cdot 10^{-4}$ per reactor year, at a subjective confidence level of 95%. In carrying out the analysis constant failure rates were assumed for describing the failure behaviour of both components and systems.

The estimate of the expected frequency of core-damage of $5 \cdot 10^{-5}$ per reactor year agrees well with the frequency of plant damage states calculated in the DRS-B (approximately $4 \cdot 10^{-5} \, a^{-1}$). The estimated upper subjective 95% confidence limit of the precursor study ($5 \cdot 10^{-4}$ per reactor year) is close to the corresponding values of the DRS-A and DRS-B ($3 \cdot 10^{-4}$ and $1 \cdot 10^{-4}$ per reactor year, respectively).

3.3   Studies for Fast Breeder
and High Temperature Reactors

3.3.1   Risk Study for the Sodium-Cooled Fast Breeder Reactor SNR-300 [42]

The purpose of the analysis of the sodium-cooled fast breeder reactor was to provide a pragmatic safety comparison between the SNR-300 [43] and a light water reactor. As far as the specific properties of the sodium-cooled fast breeder reactor permitted, the approach of the DRS-A was used.

With the SNR-300, the reactor core can be destroyed by slow meltdown or by a fast power excursion if the safety systems fail. Both phenomena are henceforth termed core destruction. The initiating events for core destruction, which were identified in the study, are shown in Fig. 3.22.

Based on the safety analyses of the licensing procedure, reliability analyses were used to estimate the expected frequencies of the initiating events for core destruction. Table 3.12 shows the results for the three most important groups. The dominant contribution stems from the initiating event "insufficient coolant flow through the core after the failure of the reactor scram" (group 1). This case was treated as the base case for the analysis of core destruction accidents. If conditions at the beginning of the core destruction were expected to be different for other initiating events, this was accounted for in the analysis.

The probability of exceeding the design value for releases of mechanical energy during core destruction is an important parameter in the risk analysis of SNR-300. In order to have a broad basis to estimate this value, expert opinions were sought about significant phenomena which may influence the course of the accident.

Table 3.13 shows the conditional probabilities of the failure of the reactor tank after core destruction. Additionally, loads, possible failure modes and conditional failure probabilities for the inner and outer containment were analysed.

Table 3.12. Frequencies of initiating events (from [42])

Initiating event group	Accident-initiating event	Annual frequency	Failed system function	Conditional failure probability	Annual frequency of initiating event
1	Transient	12	Reactor scram (mechanical)	10^{-7}	$1.2 \cdot 10^{-6}$
2	Transient	12	Scram signal Rod insertion	10^{-7} 10^{-1}	$1.2 \cdot 10^{-7}$
5	Emergency generator demanded	0.07	Active RHR Passive RHR	10^{-4} 10^{-2}	$7 \cdot 10^{-8}$
	Disturbance in main cooling system (steam generator failure)	1	Active RHR Passive RHR	$1.5 \cdot 10^{-2} \cdot 5 \cdot 10^{-4}$ $\cdot 10^{-2}$	$8 \cdot 10^{-8}$
	General RHR failure	11	Active RHR Passive RHR	$1.7 \cdot 10^{-3} \cdot 5 \cdot 10^{-4}$ $\cdot 10^{-2}$	10^{-7}
Sum 5		12			$3 \cdot 10^{-7}$

RHR: Residual heat removal

For this purpose eight reference cases were defined which comprise the possible failure combinations of the required systems. The results of the release calculations for the eight cases were combined into five release categories. They are shown in Table 3.14.

The frequencies of the release categories also incorporate external events. A significant contribution stems from severe earthquakes. If these are of very high intensity, core destruction and the failure of the containment were assumed. Therefore, such an event mainly contributes to the release categories 1 (50%) and 3 (40%), which both refer to early containment failure.

Besides releases from the reactor core, accidents involving the fuel storage pools (sodium-cooled or gas-cooled) were analysed.

Table 3.15 shows the estimated 90% confidence intervals for the frequencies of occurrence of release categories 1 to 5 (without contributions from external events).

Accident-consequence calculations were carried out for the site of the SNR-300, Kalkar (FRG). The accident-consequence model of the DRS-A was used after slight modifications. These concerned the resuspension model for humid climate, the inclusion of actinides in the ingestion pathway and new dose factors according to ICRP-30. They led to a reduction of early fatalities but an increase of late health effects compared to the values obtained with the model from the DRS-A. However, these modifications do not significantly influence the risk figures.

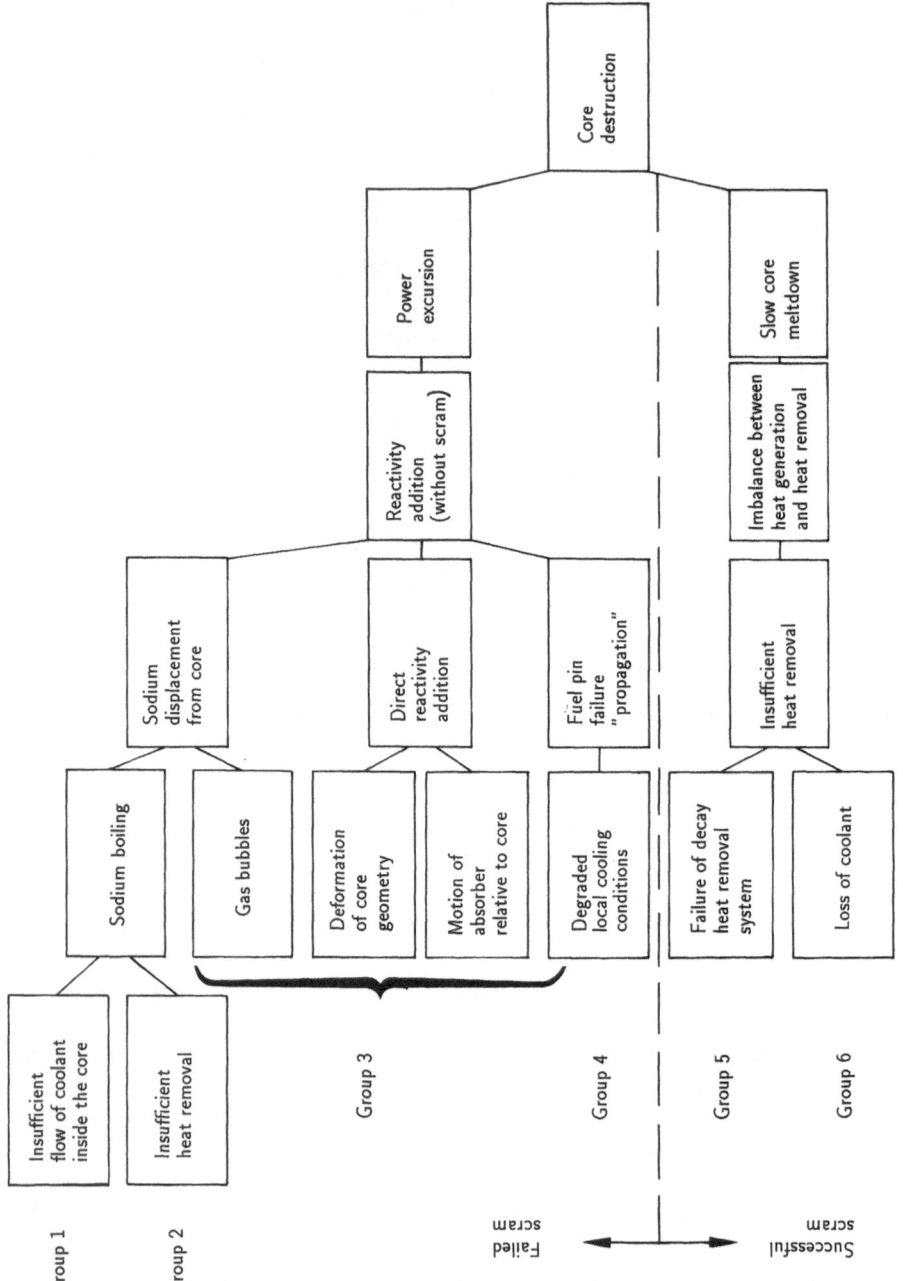

Fig. 3.22. Initiating events for the destruction of the core of the SNR-300 (from [42])

Table 3.13. Conditional probabilities of reactor tank failure after the initiating events "insufficient coolant flow after a failure of scram (UKDS)", "insufficient heat removal after a failure of scram (UWVA)", and "failure of the residual heat removal system (ANWA)" (from [42])

Initiating event	Annual frequency of occurrence	Conditional probabilities of tank failure		
		Plug system failure	Mechanical tank failure	Thermal failure (melt-through)
UKDS	$1.2 \cdot 10^{-6}$	$3 \cdot 10^{-3}$	$3 \cdot 10^{-4}$	$7 \cdot 10^{-2}$
UWVA	$1.2 \cdot 10^{-7}$	$5 \cdot 10^{-3}$	$1 \cdot 10^{-2}$	0.99
ANWA	$3.0 \cdot 10^{-7}$	–	0.5	0.5

Even for the most severe accidents the calculated radiation exposures do not cause early fatalities. This holds even if emergency measures are not taken into account.

The magnitude of late somatic health effects, i.e. the number of fatalities through cancer or leukaemia, was calculated using a linear dose-effect relationship, just as in the DRS-A. Figure 3.23 shows the results in the form of complementary probability distributions. For comparison, the results from the DRS-A, normalized to one plant are given. Similar results hold for the other damage categories.

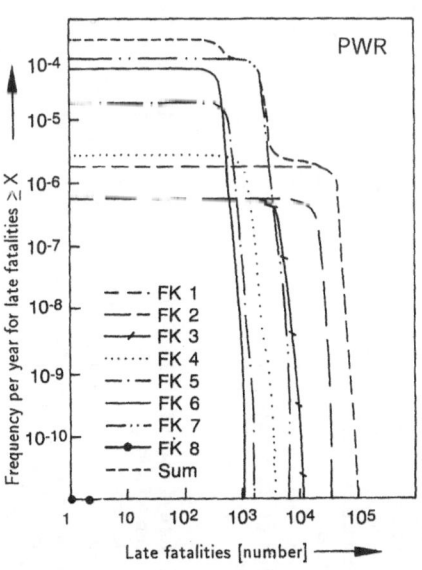 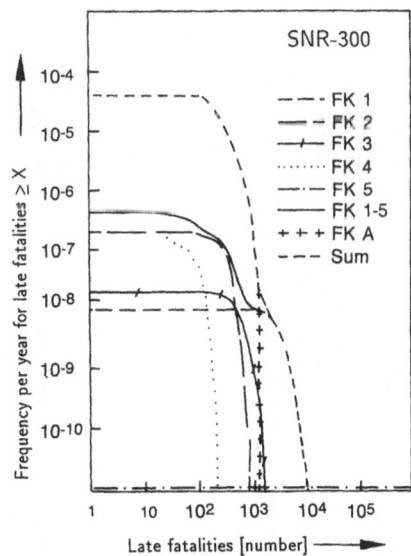

FK = Release category

Fig. 3.23. Complementary frequency distributions for late somatic effects for PWR and SNR-300 (from [42])

Table 3.14. Release categories (from [42])

Release category	Description	Time of release in h	Duration of release in h	Height of release in m	Energy released 10^6 kJ/h	Expected frequency of release in 1/year	Released fraction of core inventory Xe, Kr	I_{org}	NaI	Cs, Rb	Te, Sb[1]	Ba, Sr	La[2]
							1	–	0.15	0.15	0.15	0.05	0.05
1	Core destruction, plug syst. failure, overpressure failure of outer containment	0	1	100	530	$1 \cdot 10^{-8}$	1	–	0.15	0.15	0.15	0.05	0.05
2	Core destruction, mechanical tank failure, damaged sodium catch pan, loss of power	0	1	10	–	$2 \cdot 10^{-7}$	$5.0 \cdot 10^{-3}$	–	$2.4 \cdot 10^{-5}$	$5.2 \cdot 10^{-5}$	$7.6 \cdot 10^{-5}$	$5.4 \cdot 10^{-7}$	$2.0 \cdot 10^{-7}$
		22	1	100	15		$2.5 \cdot 10^{-1}$	–	$4.8 \cdot 10^{-4}$	$1.4 \cdot 10^{-3}$	$5.3 \cdot 10^{-3}$	$3.0 \cdot 10^{-5}$	$7.4 \cdot 10^{-7}$
		25	1	100	15		$5.7 \cdot 10^{-1}$	$2.0 \cdot 10^{-3}$	$2.0 \cdot 10^{-2}$	$1.1 \cdot 10^{-3}$	$1.4 \cdot 10^{-2}$	$4.2 \cdot 10^{-4}$	$4.0 \cdot 10^{-4}$
		33	2	100	–		$1.8 \cdot 10^{-1}$	$8.0 \cdot 10^{-3}$	$7.5 \cdot 10^{-3}$	$6.0 \cdot 10^{-5}$	$4.1 \cdot 10^{-3}$	$1.5 \cdot 10^{-4}$	$1.5 \cdot 10^{-4}$
3	Core destruction, thermal tank failure, unfiltered off-gas	0	3	100	–	$2 \cdot 10^{-8}$	$2.2 \cdot 10^{-1}$	$2.2 \cdot 10^{-3}$	$1.5 \cdot 10^{-2}$	$3.0 \cdot 10^{-2}$	$1.7 \cdot 10^{-2}$	$1.8 \cdot 10^{-4}$	$1.5 \cdot 10^{-4}$
		22	1	100	–		$2.6 \cdot 10^{-1}$	$2.6 \cdot 10^{-3}$	$6.1 \cdot 10^{-3}$	$3.0 \cdot 10^{-4}$	$5.9 \cdot 10^{-3}$	$1.2 \cdot 10^{-4}$	$1.2 \cdot 10^{-4}$
		48	1	100	–		$5.2 \cdot 10^{-1}$	$5.2 \cdot 10^{-3}$	$6.7 \cdot 10^{-3}$	–	$5.5 \cdot 10^{-3}$	$1.4 \cdot 10^{-4}$	$1.4 \cdot 10^{-4}$
4	Core destruction, thermal tank failure, loss of power, containment isolated	2	1	10	–	$2 \cdot 10^{-7}$	$7.4 \cdot 10^{-5}$	$7.4 \cdot 10^{-7}$	$8.8 \cdot 10^{-6}$	$1.6 \cdot 10^{-5}$	$7.9 \cdot 10^{-6}$	$9.3 \cdot 10^{-8}$	$9.0 \cdot 10^{-8}$
		10	1	10	–		$8.0 \cdot 10^{-4}$	$8.0 \cdot 10^{-6}$	$1.3 \cdot 10^{-5}$	$2.6 \cdot 10^{-5}$	$1.6 \cdot 10^{-5}$	$1.6 \cdot 10^{-7}$	$1.3 \cdot 10^{-7}$
		25	1	10	–		$5.5 \cdot 10^{-3}$	$5.5 \cdot 10^{-5}$	$1.2 \cdot 10^{-5}$	$3.4 \cdot 10^{-6}$	$1.2 \cdot 10^{-5}$	$2.5 \cdot 10^{-7}$	$2.4 \cdot 10^{-7}$
		48	1	10	–		$3.8 \cdot 10^{-2}$	$3.8 \cdot 10^{-4}$	$6.4 \cdot 10^{-5}$	–	$5.2 \cdot 10^{-5}$	$1.3 \cdot 10^{-6}$	$1.3 \cdot 10^{-6}$
		100	1	10	–		$9.6 \cdot 10^{-1}$	$9.6 \cdot 10^{-3}$	$7.6 \cdot 10^{-4}$	–	$6.3 \cdot 10^{-4}$	$1.6 \cdot 10^{-5}$	$1.6 \cdot 10^{-5}$
5	Core destruction, thermal tank failure	240	1	100	–	$3 \cdot 10^{-7}$	$6.2 \cdot 10^{-3}$	$6.2 \cdot 10^{-7}$	$4.6 \cdot 10^{-10}$	$8.3 \cdot 10^{-10}$	$6.0 \cdot 10^{-10}$	$5.6 \cdot 10^{-12}$	$4.4 \cdot 10^{-12}$
		280	1	100	–		$6.2 \cdot 10^{-3}$	$6.2 \cdot 10^{-7}$	$4.6 \cdot 10^{-10}$	$8.3 \cdot 10^{-10}$	$6.0 \cdot 10^{-10}$	$5.6 \cdot 10^{-12}$	$4.4 \cdot 10^{-12}$
		320	1	100	–		$6.2 \cdot 10^{-3}$	$6.2 \cdot 10^{-7}$	$4.6 \cdot 10^{-10}$	$8.3 \cdot 10^{-10}$	$6.0 \cdot 10^{-10}$	$5.6 \cdot 10^{-12}$	$4.4 \cdot 10^{-12}$
							Released fraction of spent fuel pool inventory						
A	Failure of cooling system of sodium cooled spent fuel pool	100	1	100	15	$4 \cdot 10^{-5}$	–	–	$7.3 \cdot 10^{-3}$	–	$4.4 \cdot 10^{-4}$	–	–
		102	1	100	15		–	–	$3.5 \cdot 10^{-2}$	$2.2 \cdot 10^{-2}$	$4.1 \cdot 10^{-3}$	$1.1 \cdot 10^{-9}$	$1.1 \cdot 10^{-10}$
		130	2	100	–		–	$1.0 \cdot 10^{-2}$	$9.0 \cdot 10^{-3}$	$9.0 \cdot 10^{-3}$	$1.2 \cdot 10^{-2}$	$5.7 \cdot 10^{-6}$	$5.7 \cdot 10^{-7}$
		1300	1	100	–		1.0	–	–	–	–	–	–
B	Failure of cooling system of gas cooled spent fuel pool	140	3	100	–	10^{-3}	0.5	$5.0 \cdot 10^{-6}$	$5.0 \cdot 10^{-6}$	$5.0 \cdot 10^{-6}$	$5.0 \cdot 10^{-7}$	$5.0 \cdot 10^{-10}$	–
		200	2	100	–		0.5	$5.0 \cdot 10^{-6}$	$5.0 \cdot 10^{-6}$	$5.0 \cdot 10^{-6}$	$5.0 \cdot 10^{-7}$	$5.0 \cdot 10^{-10}$	–

[1] including Se [2] including Ru, Rh, Co, Mo, Tc, Y, La, Zr, Nb, Ce, Pr, Nd, Np, Pu, Am, Cm

Table 3.15. Mean values and 90% confidence intervals of frequencies of occurrence of release categories (internal initiating events) (from [42]).

Release category	Annual frequency of occurrence		
	5% percentile	Mean	95% percentile
1	$2 \cdot 10^{-10}$	$6 \cdot 10^{-9}$	$5 \cdot 10^{-7}$
2	$2 \cdot 10^{-9}$	$1 \cdot 10^{-7}$	$5 \cdot 10^{-6}$
3	$2 \cdot 10^{-19}$	$1 \cdot 10^{-8}$	$1 \cdot 10^{-6}$
4	0.0	$1 \cdot 10^{-7}$	$3 \cdot 10^{-6}$
5	$7 \cdot 10^{-9}$	$2 \cdot 10^{-7}$	$8 \cdot 10^{-6}$

3.3.2 Risk Analyses for High Temperature Reactors

3.3.2.1 The AIPA Study

The "Accident Initiation Progression Analysis (AIPA)" study [44] was initiated in 1974 with the intention of providing guidance for research and development projects on the safety of high temperature (HTR) reactors. Furthermore the safety of alternative designs for HTRs was to be examined. It was intended to improve the methods of HTR analysis and to create the corresponding data base.

The study was published in 1976. In compliance with its objectives only a limited number of accidents, which were believed to dominate the risk, were analysed. The radiological impacts on the surroundings are given in terms of individual doses at the perimeter of a circular area with a radius of 2 km or as collective doses for this entire area. The average site data from the RSS were used for these calculations. In addition, the individual dose at the perimeter of the exclusion zone which lies at a distance of 500 m from the reactor was estimated.

Seventeen initiating events were chosen for detailed treatment. They are given in Table 3.16. After a preliminary analysis some of the initiating events were investigated in even more detail. The corresponding results are shown in Fig. 3.24.

In an evaluation of the AIPA study performed by the Nuclear Research Centre (KFA) in Jülich (FRG) it was concluded that the expected values for the risk of the dominant accident sequences were too low by a factor of up to 100. In addition, it was recommended to analyse further accident sequences.

3.3.2.2 Safety Analysis for the High Temperature Reactor Under German Siting Conditions

Scope of the Study

The risks of the High Temperature Reactor (HTR) were investigated at the Nuclear Research Centre at Jülich (FRG) [45, 46]. The reference plant of the study is the concept of the HTR-1160 nuclear power plant [47, 48]. It is a

Table 3.16. Representative initiating events (from [44])

Classes of radioactive sources	No.	Representative initiating events
Secondary coolant	7	Loss of offsite power
	11	Steam line break
Primary coolant	3	Rapid PCRV depressurization
	4	Slow PCRV depressurization
	1	Helium instrumentation line break
	2	PCVR purge header break
	9	Reheater tube leakage
Reactor core	7	Loss of offsite power
	16	Safe shutdown earthquake
	17	PCRV structural failure
	8	Moisture inleakage to coolant
Spent fuel	5	Drop of fuel shipping cask
Gaseous waste	6	Gas waste tank rupture
Liquid waste	12	Liquid waste tank leak
Solid waste	14	Drop of irradiated hardware
Irradiated material	13	Drop of control rod shipping cask
Neutron source	10	Drop of neutron source
Fuel element handling	15	Drop of recycle fuel container

PCRV: Prestressed concrete reactor vessel.

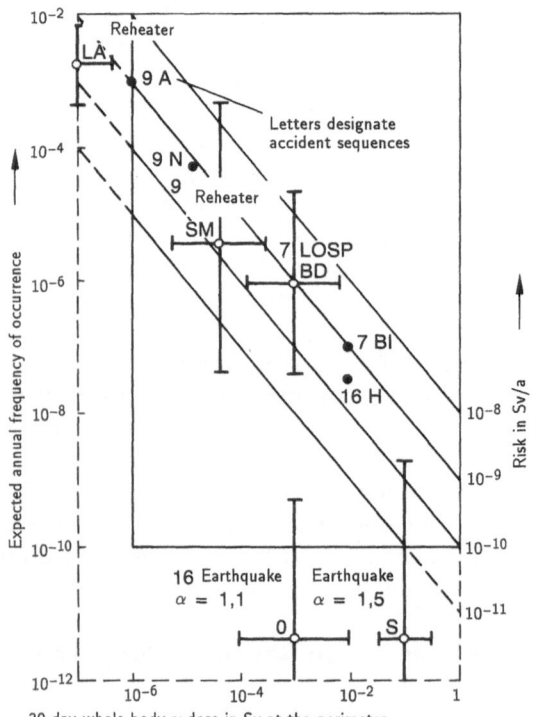

30-day whole body-γ-dose in Sv at the perimeter of the zone of low population density

α : quotient of the magnitude of the actual earthquake and that of the design earthquake

Fig. 3.24. Results of the detailed analysis of the AIPA study (from [44])

helium-cooled high temperature reactor with 3000 MW thermal power. The fuel elements are graphite blocks of hexagonal cross-section with cooling channels and axial boreholes for the fuel. The fuel rods consist of a homogeneous mixture of enriched uranium as fuel and thorium as breeding material. Both are embedded in a graphite matrix.

The nuclear steam supply system – reactor core, coolant blowers, and steam generators – is located inside the reactor pressure vessel which is constructed from prestressed concrete. The nuclear system is enclosed by a reactor building.

The analysis comprises the following three phases:

Phase Ia: Critical assessment of the AIPA study

Phase Ib: Adaptation of the results of the AIPA study to German conditions; the HTR-1160 served as a reference plant, and German site conditions were used. The method of the AIPA study was employed after eliminating some weaknesses.

Phase II: Risk analysis for a German HTR plant with high power and spherical fuel elements.

The objective of the study was to identify the largest single contributions to risk in the form of representative releases and their corresponding expected frequencies of occurrence.

For this purpose

- events were identified which could initiate a severe accident;
- possible accident sequences were determined, their risk contributions roughly estimated and their significance judged;
- events which were expected to make dominant contributions to risk were analysed in detail and the corresponding uncertainties quantified;
- events which were assumed to contribute insignificantly to risk were only roughly analysed or the bounds on their contributions were estimated.

Approach and Results

The methodology largely corresponds to that of other risk studies (cf. Chap. 2).

Common cause failures were treated by the Beta-Factor Method (cf. Sect. 2.1.4.7). In general, a β-factor of 0.1 was used. If a sufficient statistical data base was available, specific β-factors were employed: emergency diesels $\beta = 8 \cdot 10^{-3}$ per demand, $3 \cdot 10^{-2}$ during operation; for feed water pumps $\beta = 0.02$ and for components of the steam supply system $\beta = 0.2$.

Whenever possible, the failure rates and other reliability parameters of the DRS-A were used. For HTR specific components data were obtained from the operating experience with gas-cooled reactors which had accumulated in the UK until 1979.

Transients with core heat-up (with a separate treatment for station blackout) and steam generator leaks with water ingress into the primary coolant circuit turned out to be the risk-dominant events, as shown in Table 3.16.

The following initiating events of less significance were also analysed:

- leaks in the reheater,
- depressurization of the primary coolant system,
- external events (in particular, earthquakes),
- fires,
- anticipated transients without scram.

The expected frequencies of the dominant events and of some of the events of less significance were assessed using event and fault tree analyses. Essential results are failure modes, times and probabilities of failure of the prestressed concrete reactor building (RSG) and the quantities of radioactive substances released into the environment. Uncertainties of the frequencies of occurrence and of the magnitude of release were also calculated.

In the study a number of system modifications were proposed which are supposed to reduce the frequency of accidents. Examples are improvements to the auxiliary steam supply system, control of the refilling of the feed water tank from the control room, and an improved possibility of isolating a defective steam generator. Table 3.17 also shows the expected frequencies of occurrence of the release categories if these improvements are taken into account.

3.4   Risks of the Nuclear Fuel Cycle

3.4.1   Overview of the Fuel Cycle

Irrespective of the specific nuclear fuel and type of reactor, the stations of the nuclear fuel cycle comprise plants for the fuel supply (front end) and plants for waste disposal (back end). The fuel supply consists of uranium mining and preparation, the conversion of uranium to UF_6, its enrichment with the nuclide U-235 unless natural uranium is used, its conversion into an appropriate compound, e.g. UO_2, and the fabrication of fuel elements. A fuel cycle that includes reprocessing also requires interim storage of spent fuel elements, fabrication of mixed oxide fuel elements[5], waste treatment and final storage. Waste treatment includes the conditioning and the storage of waste before and after the process. The time frames and the progression of accidents with releases of radioactive substances from final storage differ largely. For this reason the operation and post-operation phases (sealed final storage) have to be distinguished. Other stations of the cycle are the transport of nuclear fuel and waste, decommissioning and dismantling of nuclear installations. The most important plants of the nuclear fuel cycle are shown in the flow chart for the radioactive materials of Fig. 3.25. If the spent fuel elements are sent directly to the final storage, fewer steps and plants are needed.

---

[5]   Mixed oxide fuel elements (MOX-FE): uranium-plutonium-oxide fuel elements.

Table 3.17. Dominant release categories and expected values for frequencies of occurrence, releases and times of release for the HTR-1160 (from [45])[1]

No.	Event, characteristic	Frequency per reactor year — original design	modified design	Time in hours	Xe-Kr	I	Te-Sb	Sr-89 Ba	Sr-90	Cs
	Accumulated release in Bq (total inventory *)				$1.1 \cdot 10^{19}$ *	$1.9 \cdot 10^{19}$ *	$3.7 \cdot 10^{18}$ *	$1.1 \cdot 10^{19}$ *	$2.6 \cdot 10^{17}$ *	$7.4 \cdot 10^{17}$ *
	Core heat-up KA-									
8	No concrete destruction RSG[2] intact	10^{-3} (6)	$2 \cdot 10^{-4}$	14	$7.4 \cdot 10^{11}$	$2.2 \cdot 10^{11}$	—	—	—	—
				720	$3.7 \cdot 10^{16}$ (5)	$1.1 \cdot 10^{14}$ (5)	$3.7 \cdot 10^{13}$ (6)	$1.1 \cdot 10^{13}$ (7)	$3.7 \cdot 10^{11}$ (7)	$1.1 \cdot 10^{13}$ (4)
7	Concrete destruction RSG-failure > 21 days	$3 \cdot 10^{-4}$ (25)	$6 \cdot 10^{-5}$	14	$7.4 \cdot 10^{11}$	$2.2 \cdot 10^{11}$	—	—	—	—
				515	$7.4 \cdot 10^{17}$ (5)	$3.7 \cdot 10^{14}$ (6)	$1.9 \cdot 10^{14}$ (9)	$7.4 \cdot 10^{14}$ (16)	$2.6 \cdot 10^{13}$ (16)	$1.9 \cdot 10^{14}$ (10)
5	Concrete destruction RSG-failure ~ 7 days	$7 \cdot 10^{-5}$ (28)	$3 \cdot 10^{-7}$	14	$7.4 \cdot 10^{11}$	$2.2 \cdot 10^{11}$	—	—	—	—
				170...250	$3.0 \cdot 10^{18}$ (4)	$3.0 \cdot 10^{16}$ (6)	$2.6 \cdot 10^{14}$ (8)	$2.6 \cdot 10^{17}$ (8)	$1.9 \cdot 10^{16}$ (8)	$1.5 \cdot 10^{14}$ (5)
4	Early concrete destruction RSG-failure ~ 4.5 days	10^{-6} (38)	$7 \cdot 10^{-9}$	14	$7.4 \cdot 10^{11}$	$2.2 \cdot 10^{11}$	—	—	—	—
				110...250	$3.7 \cdot 10^{18}$ (4)	$3.7 \cdot 10^{16}$ (5)	$3.7 \cdot 10^{15}$ (8)	$1.1 \cdot 10^{18}$ (8)	$3.0 \cdot 10^{16}$ (8)	$1.1 \cdot 10^{17}$ (5)
2	Concrete destruction no RSG isolation	$4 \cdot 10^{-7}$ (50)	$2 \cdot 10^{-8}$	14	$7.4 \cdot 10^{14}$	$7.4 \cdot 10^{14}$	$1.5 \cdot 10^{14}$	—	—	—
				120...250	$3.7 \cdot 10^{18}$ (2)	$3.7 \cdot 10^{16}$ (6)	$7.4 \cdot 10^{14}$ (6)	$7.4 \cdot 10^{16}$ (7)	$2.6 \cdot 10^{15}$ (7)	$2.2 \cdot 10^{16}$ (5)
1	Very early concrete destruction no RSG isolation	10^{-7} (19)	10^{-9}	14	$7.4 \cdot 10^{14}$	$7.4 \cdot 10^{14}$	$1.5 \cdot 10^{14}$	—	—	—
				70...250	$3.7 \cdot 10^{18}$ (2)	$1.5 \cdot 10^{17}$ (4)	$3.7 \cdot 10^{16}$ (6)	$3.7 \cdot 10^{15}$ (7)	$3.7 \cdot 10^{16}$ (5)	—
	Water ingress									
WE-1	Failure to isolate leak no containment	$8 \cdot 10^{-5}$ (27)	$4 \cdot 10^{-6}$	1.5	$7.4 \cdot 10^{14}$	$2.2 \cdot 10^{14}$	—	$3.7 \cdot 10^{12}$	$3.7 \cdot 10^{12}$	$1.5 \cdot 10^{14}$
				25	$1.5 \cdot 10^{16}$ (7)	$2.2 \cdot 10^{15}$ (6)	—	$7.4 \cdot 10^{13}$ (21)	$7.4 \cdot 10^{13}$ (26)	$2.2 \cdot 10^{15}$ (34)
	Depressurization									
1	Leak in primary circuit no RSG isolation	$3 \cdot 10^{-6}$		1	$1.9 \cdot 10^{14}$	$3.7 \cdot 10^{13}$	—	$7.4 \cdot 10^{11}$	$7.4 \cdot 10^{11}$	$3.7 \cdot 10^{11}$

* Total inventory

[1] the error factors in brackets are the square root of the quotient of the upper and lower limits of the confidence intervals

[2] RSG: prestressed concrete reactor building

The commercial utilization of large-scale plants for the nuclear fuel supply is an established technology. The safety of such plants has been proved. Waste disposal facilities, however, are still under development. Technical descriptions to the degree of detail of those for nuclear power plants are only available for the few plants of the nuclear fuel cycle which are in operation.

The investigations of the radiological risks of the nuclear fuel cycle either refer to the complete fuel cycle or they treat some of its stations. A survey is given in [50].

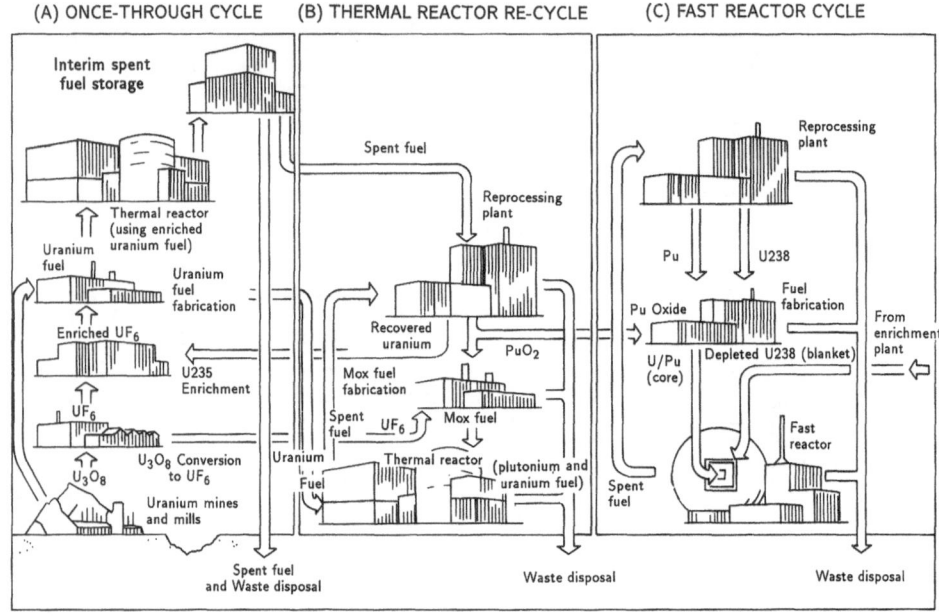

Fig. 3.25. Overview of the nuclear fuel cycle (after [49])

3.4.2   Studies for the Entire Fuel Cycle

Studies dealing with the entire fuel cycle are summarized in Table 3.18. They are carried out for estimating the radiation dose or the radiological risk and the frequency of occurrence of accidents.

Among them the study in [51] is the only probabilistic risk analysis. Its objective was to assess the risk contributions from the plants of the nuclear fuel cycle. The study is based on the methodology of the RSS. The frequencies of occurrence of accidents were calculated by fault tree analyses. In the accident-consequence model the atmospheric dispersion was treated with the

Table 3.18. Overview of risk studies for the entire fuel cycle

	[52]	[49]	[53]	[54]	[51]	[55]	[56]
Uranium mining	U/B	–	N + U/A + B	N/A + B	N/B	N/A + B	N/B
Uranium preparation	U/B	(×)	N/A + B	N/A + B	N/B	N/A + B	N/A + B
Conversion	U/B	–	N/A + B	N/A + B	–	–	N/A + B
Enrichment	U/B	×	N/A + B	N/A + B	–	–	N/A + B
FE fabrication	U/B	–	N/A + B	N + U/A + B	–	–	N/A + B
Reactor:							
Interim storage	U/B	×	–	N + U/A + B	–	–	–
Reprocessing	U/B	×	N + U/A + B	N + U/A + B	U/B	N/A + B	N/A + B
MOX-FE fabrication	U/B	×	–	N + U/A + B	U/B	×	–
Waste treatment	U/B	×	N/A + B	N + U/A + B	U/B	×	N/B
Final storage (long term)	–	×	–	U/A	U/B	×	N/B
Transport	U/B	(×)	N/A + B	N/A + B	U/B	–	N/A + B
Decommissioning	U/B	(×)	N + U/A + B	(×)	–	–	–
Remarks on the report	qualitative treatment of accidents	only some safety aspects treated	compilation of results	critical survey of literature	risk analysis for the fuel cycle	mostly technical aspects	compilation of results

– not treated
× treated globally
(×) treated partially
N normal operation; U accidents; A occupational; B population

CRAC-code which was already used in the RSS but without accounting for evacuation measures. The selected accident sequences were regarded as encompassing the spectrum of possible accidents. Here, the authors are conscious of the uncertainties of the assessment of the frequencies of occurrence and accident consequences resulting from incompleteness of the spectrum of accidents investigated, and from shortcomings of the risk model. However, they consider the results as useful estimates of the risk.

The results of the study are presented in Table 3.19. Figure 3.26 shows the risk of the individual stations of the fuel cycle, as calculated there. It contributes about 1% to the risk in nuclear electricity generation. Thus, the total risk of nuclear energy is largely determined by reactor operation.

Table 3.19. Results of the study [51]: radiological population risk (probability × magnitude of damage) of the entire fuel cycle for a production of 1 GWe a.

Station of the fuel cycle	Whole body dose in person-Sv	Late health effects
Nuclear power plant	2.57 [a]	0.02 [b]
Uranium mining and preparation		
– accident	–	–
– routine	0.002	$2 \cdot 10^{-5}$ [c]
Reprocessing	$2 \cdot 10^{-6}$	$3 \cdot 10^{-8}$
Mixed oxide fabrication	$4 \cdot 10^{-4}$	$3 \cdot 10^{-6}$
Transport	$3 \cdot 10^{-4}$	$3 \cdot 10^{-6}$
Waste storage		
– before final closure	$4 \cdot 10^{-7}$	$2 \cdot 10^{-10}$
– long term (10^6 years)	$5 \cdot 10^{-13}$	$5 \cdot 10^{-15}$ [d]
Natural radiation exposure	$7 \cdot 10^2$ [e]	

[a] Estimate from the RSS based on genetic effects according to Tables VI 9-11 and XI 4-1.
[b] From the RSS, Table XI 4-1.
[c] Based on 100 cancer fatalities per 10 000 person-Sv.
[d] Based on the 30-year commitment integrated over 10^6 years and a population of 10^6.
[e] $3 \cdot 10^8$ persons × 1.5 mSv/685 GWe a in the year 2005.

In the fuel cycle the most important contribution to risk stems from occupational accidents and diseases in uranium mining and preparation. The largest accident risk results from the transport of radioactive waste, in particular wastes containing transuranium elements, and from the fabrication of mixed oxide fuel elements (MOX-FE). Reprocessing of spent fuel elements constitutes a smaller risk. The smallest risk contribution stems from the final storage facility; here the post-operation phase is not even shown because of its extremely low risk value. The radiological risk from transport constitutes only 1/1000 of the risk of being run over by the truck carrying the spent fuel. The impacts from earthquakes were only assessed for the fabrication of MOX-FE. They make a contribution to risk which is about two orders of magnitude larger than the risk from plant internal accidents. Ignoring the contribution from earthquakes and aircraft crashes the whole body dose from the MOX-FE fabrication amounts to $2.5 \cdot 10^{-7}$ person-Sv/GWe a.

Contrary to [51], the investigation in [52] treats the safety of the plants of the nuclear fuel cycle qualitatively. The objective of the analysis was to identify areas and activities which deserve further research and development work. Release of UF_6 is considered to be the most significant accident in the front end of the fuel cycle. The radioactivity of UF_6 is insignificant; its chemical toxity constitutes the hazard. In the back end of the fuel cycle, the failure of cooling and ventilation in the storage of spent fuel elements in water pools and the storage of high active liquid waste make the most important contributions to risk. Further important accidents are criticalities, fires and explosions during reprossessing, fuel recycling and waste conditioning.

Besides the investigations listed in Table 3.18 the older analyses in [57 – 59] deserve to be mentioned. The last two studies were published at the same time as the RSS. In [57] impacts on the environment resulting from the fuel cycle with reprocessing and uranium recycling for a light water reactor with a power of 1000 MWe are modelled.

In [58], the following variants of the light water reactor fuel cycle are compared:

– direct final storage of spent fuel elements without reprocessing (one-way concept),
– reprocessing with recycling of uranium,
– reprocessing with recycling of uranium and plutonium.

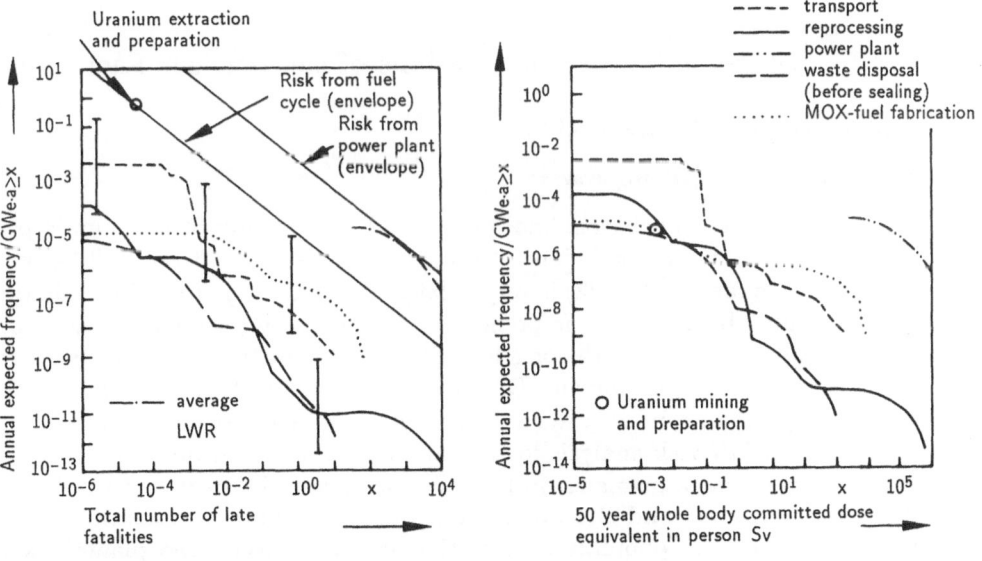

Fig. 3.26. Graphical representation of the results from [51]

The additional possibility to postpone plutonium recycling, or to finally store plutonium as waste are also examined. In the latter case, a facility for storing plutonium oxide is needed.

The report laid the foundations for numerous subsequent investigations. It mainly treats radiological impact on personnel and the general population caused by the normal operation of the plants of the fuel cycle. Accident potential is identified; in some cases individual risk at the most unfavourable location outside the plant is also estimated. The results of the study are presented in Table 3.20.

Accidents in plants of the fuel cycle for light water reactors (LWR) with reprocessing and refabrication (MOX-FE fabrication) are compiled in [59] and classified according to their risk. The stations "reactor" and "waste conditioning" are not covered. Identification of potential accidents, the corresponding frequencies of occurrence and source terms are based on the available literature. To obtain the frequencies of occurrence of accidents, statistical data from the plants, from similar chemical plants and results of event and fault tree analyses were used. The accidents were divided into five categories:

1. Explosions.
2. Fires.
3. Failures of activity barriers (pressure vessels, pipes, valves, etc.).
4. Criticalities.
5. Releases from retention equipment.

External events and those initiated by human errors were not considered. For comparison, source terms from normal plant operation and the corresponding risk are also presented.

3.4.3   Studies for Individual Stations of the Nuclear Fuel Cycle

3.4.3.1   Facilities for the Fuel Supply

Uranium mining and preparation

The radiological impact from the mining and preparation of uranium is predominantly due to the emissions of radium and radon from the mill-tailings. The noble gas radon (Rn-222) is very volatile. Therefore the resulting radiation exposure must be assessed comprehensively. The decay products of radon are high energy γ-emitters with long half lifes, a circumstance which makes resuspension processes important. Despite stabilization by protective layers, the mill-tailings are of concern even for a long time after the operation phase.

A probabilistic risk analysis for a uranium mining or uranium preparation plant has not yet been made. Published work primarily deals with radiation doses for personnel and the population during the normal operation of the plants. Risk to the population from the processes inside the plant is relatively small; therefore, the radiological impact caused by the mill-tailings is examined in most studies. In [51] the risk resulting from normal operation of a combined mining/preparation plant including waste behaviour in the post-operation phase of a model plant (stabilized mill-tailings) is assessed.

The computer code "Uranium Dispersion and Dosimetry" (UDAD) was developed for estimating the individual and population doses resulting from the process steps of a preparation plant and a mill-tailings deposit. It was used in [60], where the environmental impact of a uranium preparation plant including the radiological impact on personnel and population are examined.

Accident analyses for uranium mining and preparation facilities have not become known. External events (tornado, flooding) with the subsequent consequences for the mill-tailings as well as internal events like fire, explosion and the failure of waste-retention systems (dam rupture, leakage from pipework) are mentioned as potential causes of accidents. With a few exceptions, only qualitative assessments of accident situations can be found.

Conversion, Enrichment, Fuel Element Fabrication

The hazard of a plant for the conversion of uranium to UF_6 is chemotoxical rather than radiological. This is also true for enrichment plants, although accidents due to criticality are possible. The fabrication of UO_2-fuel elements for LWRs from enriched UF_6 is regarded as one of the safest steps of the fuel cycle, because the material to be processed is chemically stable.

In the literature the following accidents are mainly considered:

- release of UF_6 (chemotoxical risk),
- criticalities,
- fires and explosions,
- external events.

Among external events aeroplane crashes are chiefly investigated.

Criticality accidents mainly affect the plant itself; however, they can also cause significant offsite doses.

The accidental risk for these plants is estimated to be small, compared to the risks from other stations of the nuclear fuel cycle like the reactor, reprocessing, MOX-fuel element fabrication or transportation. In the probabilistic risk analysis [59] the accidental risk is estimated from statistical data on reported events in these or similar industrial plants.

3.4.3.2 Plants for the Disposal of Spent Fuel

Interim Storage of Spent Fuel Elements

The first step for disposing of spent fuel elements is their interim storage. This can be done either in the nuclear power plant itself, in external interim storage facilities or in the storage facilities of the reprocessing plants. Wet storage, that is the storage in water pools, is a proven technology for this purpose. Dry storage, especially in transport casks, has reached an advanced state of

Table 3.20. Results from [58]

Station of the nuclear fuel cycle	Whole body collective dose in person-Sv/GWe a					
	One way concept		U-recycling		Pu + U-recycling	
	Population	Occupational	Population	Occupational	Population	Occupational
Uranium mining	6.28	2.56	5.70	2.32	5.07	2.06
Uranium preparation	1.22	1.19	1.11	1.08	0.99	0.96
UF_6-conversion	0.1	$9.8 \cdot 10^{-3}$	$8.6 \cdot 10^{-2}$	$8.9 \cdot 10^{-3}$	$7.8 \cdot 10^{-2}$	$9.8 \cdot 10^{-2}$
Enrichment	$1.7 \cdot 10^{-4}$	$7.0 \cdot 10^{-3}$	$2.8 \cdot 10^{-4}$	$7.0 \cdot 10^{-3}$	$2.5 \cdot 10^{-4}$	$7.1 \cdot 10^{-3}$
UO_2 fuel element fabrication	$5.8 \cdot 10^{-3}$	0.12	$6.0 \cdot 10^{-3}$	0.12	$5.3 \cdot 10^{-3}$	0.10
MOX fuel element fabrication	0	0	0	0	$5.8 \cdot 10^{-4}$	$4.85 \cdot 10^{-2}$
LWR-energy generation	0.76	5.62	0.76	5.62	$7.7 \cdot 10^{-1}$	5.62
Interim storage	$9.2 \cdot 10^{-4}$	$3.7 \cdot 10^{-2}$	$4.4 \cdot 10^{-5}$	$1.7 \cdot 10^{-2}$	$2.6 \cdot 10^{-5}$	$1.02 \cdot 10^{-2}$
Reprocessing	0	0	3.66	$2.4 \cdot 10^{-1}$	3.54	0.26
Transport	$1.1 \cdot 10^{-3}$	$3.6 \cdot 10^{-3}$	$1.2 \cdot 10^{-3}$	$3.7 \cdot 10^{-3}$	$1.6 \cdot 10^{-3}$	$8.4 \cdot 10^{-3}$
Waste treatment	$2.2 \cdot 10^{-5}$	$3.5 \cdot 10^{-3}$	$1.8 \cdot 10^{-5}$	$3.4 \cdot 10^{-3}$	$2.1 \cdot 10^{-5}$	$3.2 \cdot 10^{-3}$
Entire industry (USA)	8.36	9.55	11.30	9.42	10.45	9.33
Total exposure	17.9		20.7		19.8	
Additional exposure from other countries	0.52		2.90		2.72	

Whole body collective dose: US industry from 1975 until 2000

	One way concept		U-recycling		Pu + U-recycling	
	Population	Occupational	Population	Occupational	Population	Occupational
Uranium mining	$2.97 \cdot 10^4$	$1.21 \cdot 10^4$	$2.65 \cdot 10^3$	$1.08 \cdot 10^4$	$2.30 \cdot 10^4$	$9.36 \cdot 10^3$
Uranium preparation	$5.79 \cdot 10^3$	$5.62 \cdot 10^3$	$5.18 \cdot 10^3$	$5.02 \cdot 10^3$	$4.49 \cdot 10^3$	$4.36 \cdot 10^3$
UF_6-conversion	$4.20 \cdot 10^2$	$4.48 \cdot 10^1$	$3.87 \cdot 10^2$	$4.00 \cdot 10^1$	$3.42 \cdot 10^2$	$3.45 \cdot 10^1$
Enrichment	$7.63 \cdot 10^{-1}$	$3.40 \cdot 10^1$	1.30	$3.50 \cdot 10^1$	1.11	$2.70 \cdot 10^1$
UO_2 fuel element fabrication	$2.51 \cdot 10^1$	$5.05 \cdot 10^2$	$2.57 \cdot 10^1$	$5.05 \cdot 10^2$	$2.23 \cdot 10^1$	$4.37 \cdot 10^2$
MOX fuel element fabrication	0	0	0	0	2.96	$2.46 \cdot 10^2$
LWR energy generation	$3.06 \cdot 10^3$	$2.27 \cdot 10^4$	$3.06 \cdot 10^3$	$2.27 \cdot 10^4$	$3.10 \cdot 10^3$	$2.27 \cdot 10^4$
Interim storage	$2.83 \cdot 10^{-1}$	$1.12 \cdot 10^2$	$1.41 \cdot 10^{-1}$	$4.00 \cdot 10^1$	$8.46 \cdot 10^{-2}$	$3.35 \cdot 10^1$
Reprocessing	0	0	$1.08 \cdot 10^4$	$7.20 \cdot 10^2$	$1.07 \cdot 10^4$	$7.74 \cdot 10^2$
Transport	$1.50 \cdot 10^1$	$5.40 \cdot 10^1$	$1.60 \cdot 10^1$	$5.60 \cdot 10^1$	$2.00 \cdot 10^1$	$8.10 \cdot 10^1$

Station	Whole body	Gastro-intestinal-tract	Bones	Liver	Kidney	Thyroid	Lung	Skin
Waste treatment	$5.90 \cdot 10^{-2}$		$6.00 \cdot 10^{1}$		$4.50 \cdot 10^{-2}$	$5.90 \cdot 10^{1}$	$5.20 \cdot 10^{-2}$	$5.20 \cdot 10^{1}$
Entire industry (USA)	$3.90 \cdot 10^{4}$		$4.11 \cdot 10^{4}$		$4.60 \cdot 10^{4}$	$3.99 \cdot 10^{4}$	$4.17 \cdot 10^{4}$	$3.81 \cdot 10^{4}$
Total exposure	$8.01 \cdot 10^{4}$				$8.59 \cdot 10^{4}$		$7.98 \cdot 10^{3}$	
Additional exposure from other countries	$2.1 \cdot 10^{3}$				$9.12 \cdot 10^{3}$		$8.90 \cdot 10^{3}$	

Ratio of the occupational and population collective doses for the LWR industry in the US (1975–2000) (only Pu + U-recycling)

Station of the nuclear fuel cycle	Whole body	Gastro-intestinal-tract	Bones	Liver	Kidney	Thyroid	Lung	Skin
Uranium mining	0.41	10	0.18	0.50	0.15	165	7.1	165
Uranium preparation	0.97	9.4	1.2	0.47	0.12	137	28	137
UF$_6$-conversion	0.10	0.49	0.52	8.4	1.2	95	35	330
Enrichment	29	0.22	270	13	2.3	170	140	90
UO$_2$ fuel element fabrication	20	18	1.6	9500	7.9	11000	13000	11000
MOX fuel element fabrication	83	1400	1.8	18	20	4400	92	4400
LWR energy generation	7.3	7.3	2.1	7.2	7.4	4.6	7.4	7.2
Interim storage	400	400	400	400	400	400	180	4.7
Reprocessing	0.14	0.094	0.058	0.14	0.14	0.079	0.13	0.024
Transport	5.2	5.2	5.2	5.2	5.2	5.2	5.2	5.2
Waste treatment	164	110	17	87	110	280	220	280
Entire industry (USA)	0.93	1.8	0.44	1.0	0.34	1.5	5.7	0.54

development. It is particularly suitable for long-term interim storage, because it does not require sophisticated safety systems. For example, passive cooling by natural convection is sufficient.

Most risk-oriented investigations for interim storage facilities of the nuclear fuel cycle refer to wet storage; for dry storage only comparative analyses were performed. A comprehensive study is presented in [61] where alternative possibilities for handling and storing spent nuclear fuel are examined. The results of accident analyses for interim storage facilities in Finland and Sweden are described in [62]. A probabilistic risk-oriented evaluation of potential accidents has been performed within the framework of the German "Projekt Sichere Entsorgung" (PSE) [63] for the concept of a nuclear disposal centre (NEZ) in Gorleben (Status: 1977). The quantities of released radioactive substances and the expected frequencies of releases were assessed.[6]

The investigations of accidents in wet storage facilities are concerned with

- the failure of the cooling system,
- leakage from the storage pool or from auxiliary systems,
- low water level (reduced shielding) or insufficient water purification,
- criticalities,
- leaks from fuel rods (defective fuel rod cladding),
- dropping of a transport cask for fuel elements (handling accidents),
- external events (tornado, earthquake).

Slowly progressing accidents are considered, in which the fuel claddings lose their integrity due to overheating of the fuel elements, and rapid events accompanied by an immediate release of radioactive substances due to massive mechanical impacts, such as the dropping of fuel elements due to faulty handling or external events. The complete loss of the storage pool water, for instance during an earthquake, is regarded as the most severe accident in an interim storage facility.

In the case of slowly progressing accident sequences sufficient time is normally available for counter-measures after the safety systems have failed. The active cooling of a wet storage pool can be interrupted without danger for a relatively long time. Hence, the risk from such failures is insignificant. On the other hand, handling accidents during the unloading of fuel elements are important. For dry storage, only handling accidents or seal failures are important.

Reprocessing Plants

Reprocessing plants consist of an entrance storage facility, the head end for the mechanical shredding of the fuel elements and the facility for the chemical dissolution and separation of reusable fuel from radioactive waste. The PUREX process [64] is the most widely used for this. Finally, there is the tail end facility

---

[6]   In the risk analysis for the entire fuel cycle including reprocessing and recycling [51] there is no need to consider interim storage, since the spent fuel elements are transported to the reprocessing plant after 90 days of decay for being reprocessed there.

for the processing of the end products uranium and plutonium. Additionally, there are storage facilities for the different types of radioactive waste. Depending on the type of plant and the fuel cycle, there may be complex systems for the purification of exhaust gases and retention equipment. The tail end consists of plants for the conversion of uranium and plutonium into products like uranium oxide or uranium hexafluoride and plutonium oxide or plutonium nitrate which are processed in other plants of the fuel cycle. Additionally, there are installations for waste conditioning. An important contribution to risk stems from the facilities for solidification of high active waste. In all parts of the plant the material to be processed is present in easily dispersable form. The confinement of radioactive material is therefore important, particularly in the case of accidents.

During the normal operation of the plant the radiological impact on the environment mainly results from the release of the nuclides H-3, C-14, Kr-85, and I-131, which escape primarily during the shredding of fuel elements. In accident situations the source term additionally contains further fission products and actinides (α- and β-emitting aerosols). In normal operation discharges to the atmosphere and to rivers are considered. Usually only the risks of releases into the atmosphere are analysed. According to [65], a risk analysis of an integrated waste disposal centre has to treat releases into the atmosphere and rivers during normal operation and accident situations. Therefore the accident-consequence models must be extended accordingly (cf. Sect. 3.1.2).

In [51, 59, 63, 66, 67] the population risks from accidents in reprocessing plants are assessed. Occupational risk from accidents and normal operation is addressed in [67]. The investigations in [51] and [67] are very similar in their approach to identifying the relevant accidents (Preliminary Hazard Analysis) and assessing the accident consequences (CRAC code). In general, fault tree analysis is used in these studies to estimate the frequencies of accidents. In [65] the fault trees are constructed using the "leak-path method" which traces the path of the released radionuclides in the direction opposite to their migration through the barriers.

The most important accidents in reprocessing plants stem from the use of flammable materials, the properties of the nuclear fuel and the high activities of the radioactive substances present in the plant and in the associated waste storage facilities. Significant accidents are:

- fire, mainly of solvents containing radioactive materials;
- explosions (red oil, H_2, solvents containing nitrates);
- criticality;
- leakage and transfer accidents;
- failure to cool self-heating fluids;
- failure of the air conditioning system.

External events are rarely analysed, since it is assumed that reprocessing plants are designed to cope with them.

The following accidents make the main contributions to risk:

- failure of the Kr-storage bottle;
- explosion of the calcinator for high radioactive waste;
- dropping of a fuel element.

The accident "failure of the Kr-storage bottle" is assumed to encompass Zr-fires and accidents with uncontrolled releases of volatile and gaseous fission products from the exhaust system (Kr-recovery, iodine-scrubber, Ru-absorber).

According to the probabilistic risk analysis [67] the risk of a reprocessing plant stems primarily from its normal operation (cf. Table 3.21).

In the FRG comprehensive investigations on the risks of reprocessing plants were carried out within the framework of PSE [63]. The plant concept analysed comprises a storage for spent fuel elements using transport casks, a hot cell type reprocessing plant and a vitrification plant based on the PAMELA procedure. Individual and population doses from important internal initating events were assessed. In the context of the project PAE [68] the results and procedures from PSE were applied to a model plant using remotely controlled modules. Figure 3.27 shows the resulting population doses.

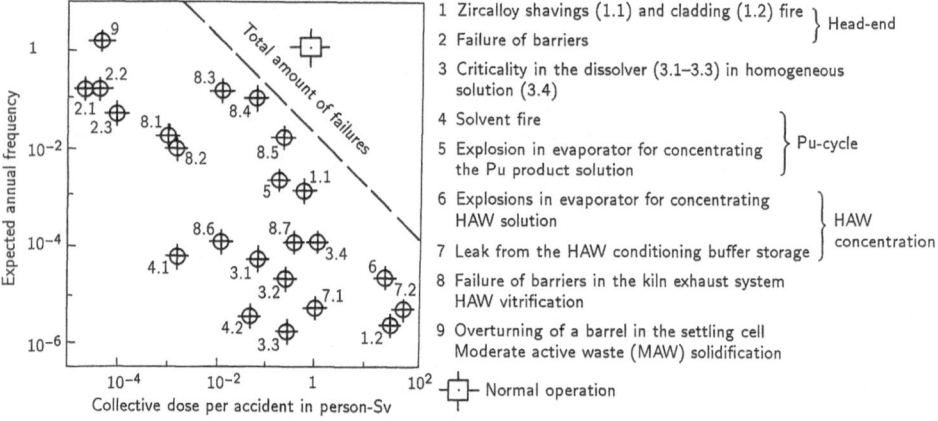

Fig. 3.27. Collective doses from accidents in reprocessing and waste treatment (according to [68])

The radiation exposure from accidents lies below that from the normal operation of the reprocessing plant. The expected values of radiation exposure for important accidents like evaporator explosion, zircalloy cladding fire or criticality were found to be approximately four orders of magnitude lower than those from natural and man-made radiation exposure.

Mixed Oxide Fuel Element Fabrication

Fuel cycles with recycling of uranium and plutonium require plants for the fabrication of mixed oxide fuel elements (MOX-FE). Differences in radiological impacts – compared to plants for the fabrication of UO_2-fuel elements – result from the use of plutonium. The staggered barriers and the air exhaust system are of importance for the analysis of accidents. As long as the filters are intact, impact is restricted to the plant and its personnel. Offsite impact varies considerably with the retention capability of the filters, which constitute the last barrier against release into the atmosphere.

Probabilistic risk analyses for a MOX-fuel element fabrication plant were performed in [51, 67, 69, 70]. The results of [67] for the radiological risk are shown in Table 3.21. According to this analysis, the threshold for early fatalities is not reached. In [69], the expected values of the individual doses were estimated for a number of accidents. The results are given in Table 3.22.

Table 3.21. Results of the risk study in [67]: Number of late fatalities per plant year for a reprocessing plant and mixed oxide fabrication plant. (Radiation doses are below the threshold for early fatalities).

Risk $\left[\dfrac{\text{late cancer fatalities}}{\text{plant year}}\right]$	FRP (PT/Ref.)[a]	FRP-WTF	FFP (PT/Ref.)[a]	FFP-WTF
Accidents				
– population	$7 \cdot 10^{-7}$	$8 \cdot 10^{-9}$	$5 \cdot 10^{-8}/1.5 \cdot 10^{-8}$	$6 \cdot 10^{-9}$
– occupational	$1.3 \cdot 10^{-6}$	$1.3 \cdot 10^{-6}$	$4 \cdot 10^{-6}$	$7 \cdot 10^{-7}$
Normal operation				
– population	$0.29/0.12$	0.24	$6.8 \cdot 10^{-4}/1.9 \cdot 10^{-4}$	0.12
– occupational	$4.5 \cdot 10^{-4}$	$4.5 \cdot 10^{-4}$	$4.7 \cdot 10^{-4}$	$1.8 \cdot 10^{-4}$

Results of the investigation by Fullwood (radiological risks only):
[a] If results for the two investigated fuel cycles differ, the first number denotes the PT-fuel cycle and the second the reference fuel cycle
FRP: Fuel recovery plant.
FFP: Fuel fabrication, MOX fuel elements.
WTF: Waste treatment facility.
PT: Partitioning-transmutation fuel cycle.
Ref: Reference fuel cycle.

The frequencies of occurrence of events in [69] are estimated from operating experience and data from the literature. It was the objective of the analysis to provide comprehensive information as a basis for safety analyses and the determination of siting criteria. In the studies of [51] and [67] accidents were quantified with fault tree analyses. Both studies refer to the "Westinghouse Fuels Recycle Plant" and use the same methodology.

Accidents mainly in connection with the processing of plutonium in powder form or as a solution are considered. Among them are:

Table 3.22. Results of the study [69]

Plant condition	Source term in Bq	Radioactive substance	Critical organ	Dose at a distance of 100 m in mSv [c]	Expected annual frequency of release	Annual risk in mSv [e]
Normal operation	$1.9 \cdot 10^5$ (L)[a]	Plutonium mixture[b]	Bones	$4.0 \cdot 10^{-2}$/50 years[d]	1	$4.0 \cdot 10^{-2}$
	(N)		Lung	$6.0 \cdot 10^{-4}$/50 years	1	$6.0 \cdot 10^{-4}$
Criticality	$1.1 \cdot 10^6$ (L)	Plutonium mixture	Bones	$2.8 \cdot 10^{-1}$/50 years	$8.6 \cdot 10^{-3}$	$2.4 \cdot 10^{-3}$
		Direct radiation	Whole body	$1.6 \cdot 10^{-1}$	$8.6 \cdot 10^{-3}$	$1.4 \cdot 10^{-3}$
		Noble gases and halogens	Whole body	$3.1 \cdot 10^{-1}$	$8.6 \cdot 10^{-3}$	$2.7 \cdot 10^{-3}$
		Iodine	Thyroid	2	$8.6 \cdot 10^{-3}$	$1.7 \cdot 10^{-2}$
Explosion	$3.7 \cdot 10^8$ (L)	Plutonium mixture	Bones	$9.4 \cdot 10^1$/50 years	$\sim 10^{-3}$	$9.4 \cdot 10^{-2}$
Fire	$4.1 \cdot 10^8$ (L)	Plutonium mixture	Lung	$7.0 \cdot 10^{-1}$/2 years	$< 10^{-2}$	$< 7.0 \cdot 10^{-3}$
Damaged glove box	$3.7 \cdot 10^8$ (N)	Plutonium mixture	Lung	$6.0 \cdot 10^{-1}$/2 years	$< 10^{-2}$	$< 6.0 \cdot 10^{-3}$
Fire in the fabrication	$3.7 \cdot 10^8$ (N)	Plutonium mixture	Lung	$6.0 \cdot 10^{-1}$/2 years	$2.0 \cdot 10^{-4}$	$1.0 \cdot 10^{-4}$
Fire in the ion exchanger column	$1.1 \cdot 10^6$ (L)	Plutonium mixture	Bones	$2.8 \cdot 10^{-1}$/50 years	$< 1.0 \cdot 10^{-1}$	$< 2.8 \cdot 10^{-2}$
Fire in the sintering kiln	$3.7 \cdot 10^5$ (N)	Plutonium mixture	Lung	$6.0 \cdot 10^{-4}$/year	$< 5.0 \cdot 10^{-2}$	$\ll 1.0 \cdot 10^{-4}$

[a] L = soluble, N = non soluble.

[b] Plutonium mixture = reference mixture of plutonium and americium.

[c] Estimated using the ICRP Pub 2 lung model (ILM).

[d] $4 \cdot 10^{-2}$/50 years means a dose of $4 \cdot 10^{-2}$ mSv is received over 50 years.

[e] The probable annual dose received by an individual. It may be received instantaneously or over a prolonged period as in the case of the bones and the lung.

– explosion,
– fire,
– criticality,
– failure of activity confinement,
– earthquakes, aeroplane crash.

The accidents with the largest contribution to risk for the population are:

– earthquakes with higher intensities than the design basis earthquake,
– aeroplane crash,
– criticality.

In study [51] the risk is practically determined by the accident "Earthquake with intensity exceeding the design basis".

Waste Treatment

In the following the treatment of waste resulting from the disposal of the fuel is mainly considered. The tailings associated with uranium mining and preparation, which constitute the main sources of radioactivity in the supply part, were already discussed in Sect. 3.4.3.1.

"Treatment of waste" implies all activities related to processing and conditioning, i.e. concentration, solidification and stabilization including the storage of radioactive materials which are prepared for final storage. Basically, the areas "storage" and "solidification" must be distinguished.

Waste is classified into several categories: HAW (high active waste), MAW (moderate active waste) and LAW (low active waste), according to such criteria as specific activity or surface doses. The classification is not internationally uniform.[7] In fuel cycles with reprocessing, HAW is defined as the waste from the first dissolution step for the separation of spent fuel elements in reprocessing plants [55]. For direct final storage all spent fuel is classified as HAW. TRU-waste (transuranium) is a special kind of waste. Its specific activity is lower than that of HAW; however, due to its α-emitting nuclides with long half lifes it requires final storage conditions similar to HAW.

HAW is conditioned for final storage by concentration/calcination and transformation into vitrified oxide form. MAW and LAW are concentrated and bound to bitumen or concrete. No standard conditioning procedure is available for the direct final storage of spent fuel elements. The Kr-85, which is generated during the reprocessing, has to be stored for a long time in metal bottles if it is extracted from the exhaust gases.

Risk estimates for storage and solidification are presented in the studies [51, 63, 67, 72]. They treat waste reprocessing (HAW, TRU, MAW, LAW, Kr-85).

---

[7]   e.g. for liquid waste in the FRG [71]
HAW: specific activity $> 3.7 \cdot 10^9$ Bq/m^3 (10^{-1} Ci/m^3)
LAW: specific activity $> 3.7 \cdot 10^6$ Bq/m^3 (10^{-4} Ci/m^3)
MAW: intermediate activity values
(Bq: Becquerel; Ci: Curie)

In [73] the storage of all kinds of waste generated in the commercial use of nuclear energy is analysed including alternative concepts for final storage.

Accidents in the storage of liquid high active waste mainly result from failures of cooling or activity confinement. For the storage of solidified high active waste failure of cooling is considered an important accident. With storage facilities for Kr-85 failures of the storage container due to corrosion, defective material or mechanical impact during handling are considered.

In facilities for the solidification of HAW and MAW failures of the devices for retaining radioactive gases are important. However, the contribution of explosions in the processing facilities of such plants to risk is negligible.

Accidents in the storage of other kinds of waste (conditioned fuel elements, TRU, MAW, LAW, Pu), and in processes like the incineration of waste or packing of spent fuel elements are rarely addressed (e.g. [73]).

Final Storage (Operation Phase)

During the operation phase of a final storage facility the materials to be stored are unloaded in a surface plant. From there they are transported to an underground storage facility. The operation phase ends with the sealing of the final storage facility. Possible accidents are uncontrolled releases of radionuclides to the biosphere.

Risk analyses for the operation phase of final storage facilities are described in [51, 73]. In both studies a bedded salt formation is considered as the final storage location. Waste from reprocessing is dealt with in [51] and spent fuel elements are treated in [73]. In [74] radiation doses for both personnel and population during normal operation are assessed. The study covers unloading of the waste, its handling and transport to the final storage. Frequencies of occurrence of accidents are also estimated.

In general, the following accidents during the operation phase of the final storage facility are addressed:

- dropping (handling accidents, drop into the pit),
- vehicle collisions (in the surface reception facility and in the underground final storage facility),
- explosions and fires,
- flooding,
- aeroplane crash,
- earthquakes,
- tornadoes,
- loss of offsite power,
- leakages,
- component and human failure.

Of particular importance are:

- dropping into the pit of the final storage facility during the storage of high active wastes (effects on personnel),
- failure of filters of the pit-ventilation system,
- crash of an aeroplane on the reception facility or close to the pit.

Long-Term Behaviour of Finally Stored Radioactive Waste

To assess potential hazards from a closed and sealed final storage facility, only accident-consequence calculations were performed. A complete risk analysis is not available due to methodological difficulties.

Geological formations like salt, granite, basalt, slate and clay are considered as possible sites for final storage. Alternatives are discharge into the sea, disposal beneath the seabed or under ice sheets, rocketing into space etc. (cf. [73]). Mostly, however, salt domes or bedded salt formations under general site conditions are analysed. The studies in [75 – 77], for example, treat existing sites.

Accidents in sealed final storage facilities can be caused by

- natural events (e.g. earthquakes, volcanic eruptions, meteorite impact),
- natural processes (e.g. climate change, erosion, sedimentation)
- human intrusions (e.g drilling, geothermal energy exploitation),
- events caused by the disposed waste itself (e.g. radiolysis, nuclear criticality, thermal expansion).

Due to the large number of conceivable combinations and sequences of accidents a comprehensive treatment of the spectrum is extremely difficult. The available studies mainly refer to the waste categories HAW, TRU, and spent fuel elements [78 – 80]. The principal investigations are discussed in the summary report [78]. Frequencies of the occurrence of natural events and events caused by man are listed. The results of accident-consequence calculations from 17 studies are compiled. For the direct final storage of spent fuel elements and for the final storage of radioactive waste the maximum individual doses lie in the range from 10^{-9} to $2 \cdot 10^{-3}$ Sv/a equivalent dose for the thyroid, the bone marrow and the intestinal tract. The figures refer to a quantity of waste corresponding to the generation of 1000 GWe a. Radiologically important are the nuclides Tc-99, I-129, Ra-226, and Np-237.

Doses were calculated for the ground water exposure pathway, which is considered to be the most probable. Releases caused by natural events like the impact of a meteorite, volcanic activities, etc. are regarded as extremely unlikely; release through human activities like deep hole drillings and nuclear explosions is not assessed.

Studies for the final storage of low active or moderate active waste have not reached the same level of quality as the analysis for high active waste. Estimates of the long-term behaviour of finally stored waste are partly contained in the work on high active waste. Some studies are only preliminary investigations;

for example, the investigation in support of the licensing procedure for the iron
ore mine Konrad in the FRG [81].

3.4.4 Transportation

Essential characteristics to be considered in the transport of radioactive materials are the motion of the radiation source and the variation of the population
distribution along the route. The potential magnitude of damage in transport
accidents is chiefly determined by the strength of the transport cask. Official
road accident statistics are used to classify the accidents according to velocities and other important criteria, and serve to estimate the frequencies of
various accident categories. Experimental investigations and theoretical analyses of damage types were mainly carried out by Sandia Laboratories [82,
83]. They provide estimates of the conditional probabilities of occurrence of
specific accident sequences resulting from fire, impact, deformation and human errors in packing or quality assurance. In the case of accidents involving
fires, the accident-consequence model must account for the thermal lift of possibly released gases or aerosols. Inhalation is the most important possibility
for incorporating radioactive substances. Thus, aerosol distributions of volatile
materials considerably influence the magnitude of damage caused by accidents.

Major studies on transportation risks are the investigations in references
[84 – 87]. Probabilistic analyses of transportation risks in the waste disposal
part of the fuel cycle are contained in references [51, 63, 88]. Further studies
are documented in [84]. The investigations performed within the framework of
the PNL-series "Transportation-Safety-Studies" are described in references [90,
91]. The results from [91] are summarized in Table 3.23. The transport risks of
plutonium on roads, railroads and by air, which were also addressed in [91], are
not shown there. The results are based on the transportation volume expected
in the US for an assumed energy generation of 100 GWe a.

Table 3.23. Results of PNL-Investigations (according to [91])

Fuel cycle step	Material shipped	Transport by	Accident risk (fatalities year^{-1})
Mining/milling	U_3O_8 (yellowcake)	Road	$< 10^{-6}$
Conversion/enrichment	UF_6	Road	$6 \cdot 10^{-5}$
Fuel fabrication	Fresh fuel	Road	$< 10^{-6}$
Power production	Spent fuel	Road, rail	10^{-5}
	Reactor wastes	Road	$2 \cdot 10^{-6}$
Fuel storage/disposal	Spent fuel	Rail	$2 \cdot 10^{-5}$
Waste management	Other fuel cycle wastes	Rail	$< 10^{-6}$
Total			10^{-4}

In the project PSE [63] the transport of spent fuel elements, vitrified HAW (glass ingots), solutions of uranyl nitrate, plutonium oxide and LAW in the FRG was mainly analysed. The population dose resulting from transport without any accidents amounts to 0.6 to 2 person-Sv a^{-1}. The transport of spent fuel elements contributes about 1/1000 to the exposure; the major contribution is due to the transport of waste. The population dose caused by accidents is assessed to be less than 1% of the above values.

3.4.5   Decommissioning of Nuclear Installations

Decommissioning comprises all activities which lower the radiation dose caused by the radioactive inventory of the plant at its site to such an extent that an unrestricted clearance of the plant and its site becomes possible. A risk analysis on the subject has not been published. Investigations in this field concern:

– reactors (BWR, PWR),
– reprocessing plants,
– fuel element fabrication (U-FE, MOX-FE),
– shallow land geological LAW-storage.

Existing reference plants which are considered as representative of the state of the art provide the basis for the analyses. Radiation doses for personnel and population due to routine work and accidents are to be estimated.

3.4.6   Summary and Outlook

Risk-oriented analyses of the individual stages of the fuel cycle vary considerably with regard to the number of investigations and their depth of analysis. The majority of the studies are concerned with accident consequences for the population.

Only a few risk studies have been carried out on plants for the fuel supply, whereas there have been several investigations into the disposal of spent fuel. The degree of detail of the risk studies depends on the maturity of the technology considered and on the interrelationship with existing plants.

In principle risk analyses are suited for the estimation of the risk from operation of the plants of the fuel cycle. Reprocessing plants are frequently treated because this technology is well established, so that plants are investigated which have been in operation for many years. However, the impact of external events is seldom taken into account. Risk analyses for the transportation of radioactive materials are also relatively frequent. Applications of risk analysis to the long-term behaviour of finally stored radioactive waste in various geological formations meet restrictions. No generally accepted methodology is presently available for the probabilistic investigation of scenarios for the penetration of radionuclides into the ground water. The identification and description of dominant accident sequences is based on an engineering appraisal of the maximum environmental impacts.

3.5 Conclusions from Risk Studies for Nuclear Reactors

3.5.1 Generalities

Considerable qualitative and quantitative insight accrues from determination of the risk of a nuclear power plant, much of which can be put into practice.

Qualitative insight stems from the identification of those plant components, protection and safety devices, and modes of operation which contribute to risk. The subsequent quantitative evaluation then shows the risk-dominant accident sequences. It also shows how risk is diminished if the frequencies of occurrence of such sequences are reduced. Such investigations are of prime interest when judging whether proposed measures for plant improvement are adequate and balanced. In plant-specific risk analyses, insight into peculiarities of plant design and modes of operation may be of interest even if they are not related to risk-dominant accident sequences.

Although large uncertainties still remain with risk analyses of nuclear installations, many useful results are obtained which are not solely derived from or are dependent upon the calculated numbers. For practical purposes the most important result of a risk analysis is the comprehensive description and documentation of design characteristics, generated in the modelling of accident sequences. The determination of expected frequencies of accident sequences does not have to be more accurate than is required for distinguishing between risk-dominant and less important sequences. The insights into plant design and modes of operation can be taken into account in the licensing procedure. Further development of licensing requirements may benefit from them (cf. Chap. 6).

Determination of the relative importance of risk-dominant accident sequences allows cost-benefit analyses for potential plant modifications to be carried out, both with regard to increased plant safety and to a balanced design of the plant.

The models developed in risk analysis can be used to optimize test and maintenance intervals and to improve modes of operation and personnel training. Insights from uncertainty analyses can help determine research priorities, particularly if large uncertainties are shown to exist in risk-dominant accident sequences.

Investigations of nuclear power plants have revealed that the essential contribution to severe environmental impact results from core destruction accidents and that the integrity of the containment is of great importance for the prevention and mitigation of offsite consequences.

The analyses have shown that the most important contributions to uncertainties in the results occur in the treatment of the following topics:

Level 1:

Treatment of common cause failures and human behaviour as well as the modelling of physical processes, particularly of thermal-hydraulic phenomena in emergency core cooling and transients.

Level 2:

Modelling of the physical phenomena during core destruction as well as modelling of physical and chemical phenomena associated with the transport and deposition of radioactive substances within the containment; determination of the time and mode of containment failure and the treatment of hydrogen explosions.

Level 3:

Selection of parameters for the atmospheric dispersion, the estimation of dry and wet deposition rates, of the thermal lift, and of the dose-effect relationships for early and late fatalities.

A comprehensive overview of insights from probabilistic investigations is given in [1, 2, 10].

3.5.2 Influence on Research Programmes

Research programmes inspired by probabilistic risk analyses are mainly concerned with accident sequences which cause severe impact on the environment, such as:

- steam explosions,
- hydrogen deflagration and detonation,
- reactor pressure vessel failure under high system pressure
- core-concrete interaction,
- chemical reactions between released radioactive substances,
- aerosol transport and deposition phenomena within the containment.

In the past, uncertainties about these phenomena have been accounted for by pessimistic assumptions. In view of the great importance of the containment and its failure modes which has been made evident in risk analyses, research programmes have been initiated for improving the understanding of such phenomena. Investigations related to steam explosions have been conducted primarily in the US, at Sandia Laboratories [92]. They led to more differentiated results which permitted a reduction in the conditional probability of the occurrence of steam explosions capable of destroying the containment during a core-melt accident in a light water reactor.

Investigations of the hydrogen problem have been performed at several institutions [93, 94].

Great importance is attached to the time history of the pressure in the containment; it is strongly influenced by the interaction of molten core material with the concrete of the basemat of the containment. At Kernforschungszentrum Karlsruhe (KFK) and at Kraftwerkunion (KWU) extensive investigations have been performed about the split-up of thermal energy into sump-water evaporation and the melting of the concrete. These investigations show that a larger part of the energy is absorbed by the melting of concrete than had earlier been assumed, as for example in the DRS-A. This leads to a slower pressure build-up in the containment due to the evaporation of water, but to higher hydrogen generation rates.

Many results of research programmes suggest that containment integrity may be preserved much longer than assumed in the RSS and the DRS-A if the containment does not fail early as a consequence of a steam or hydrogen explosion or a pressure vessel failure under high system pressure.

This increases the significance of aerosol transport and deposition behaviour and of the modelling of chemical reactions of fission products. Investigations related to this subject have been conducted at Kernforschungszentrum Karlsruhe. There are indications that the combination of longer lasting containment integrity, and of improved knowledge about aerosol behaviour will lead to a reduction of the calculated aerosol concentrations in the containment atmosphere at the time of containment failure for many accident sequences.

On the other hand, increased importance is attached to measures of hydrogen control, which are being intensely investigated [6]. As far as the results of the research programmes mentioned are sufficiently validated they have been taken into account, for example, in the DRS-B, in order to arrive at more realistic risk estimates [95 – 97].

Recent risk analyses [1, 2] have shown a considerable potential for accident management measures capable of returning the plant to a safe state after failures of safety systems in the course of an accident. The necessary procedural and hardware changes are being implemented in many nuclear power stations. In the field of mitigating measures, the ministries of research (BMFT) and environment (BMU) in the FRG for example sponsor investigations into measures for delaying or completely avoiding containment failure; as, for instance, into depressurization of the containment through controlled venting or measures for the control of hydrogen in the containment.

3.5.3 Influence on Plant Design and Modes of Operation

Risk studies reveal weaknesses of the plant design, and point to areas where the investment into safety is inadequately high. Possible risk contributions from inadequate operational procedures are also recognized. Relative contributions to risk can be estimated from the numerical results. Therefore improvements and modifications can be introduced where they are most effective. Insights obtained in the DRS-A have led to numerous improvements of the plant design and to optimized maintenance strategies. For example, manual actions of the operating crew have been partially automatized (e.g. manual cool-down of the plant at the rate of 100 K/h in the case of small leaks in the primary coolant system). Also, the possibility of recovery of offsite power after a failure to start the emergency diesels was provided. Results of risk analyses have led to a more balanced design of systems.

A comprehensive summary of system modifications in US light water reactors originating from results of probabilistic analyses of protection and safety systems can be found in [1] and [98]. In [2] the system and operational changes inspired by the investigations of the DRS-A and DRS-B are described.

References

1. Severe accident risks: An assessment for five US nuclear power plants. Summary report – second draft for peer review. Vols. 1 and 2. NUREG-1150, June 1989
2. Der Bundesminister für Forschung und Technologie (Hrsg.): Deutsche Risikostudie Kernkraftwerke – Phase B. Köln, 1990 (English Summary: German risk study nuclear power plants, phase B. GRS-74, Köln 1990)
3. Hennies, H.M.; Hosemann, J.P.: Ablauf und Konsequenzen eines DWR Kernschmelzunfalls. atw 3 (1981) 168-175
4. An assessment of steam explosion-induced containment failure
   Part I: Probabilistic aspects (T.G. Theofanous et al.). Nuclear Science and Engineering 97 (1987) 259-281
   Part II: Premixing limits (M.A. Abolfadl, T.G. Theofanous). Nucl. Science and Engineering 97 (1987) 282-295
   Part III: Expansion and energy partition (W.H. Amarasooriya, T.G. Theofanous). Nuclear Science and Engineering 97 (1987) 296-315
   Part IV: Impact mechanisms, dissipation, and vessel head failure (G.E. Lucas et al.). Nuclear Science and Engineering 97 (1987) 316-326
5. Reimann, M.; Murfin, W.: The WECHSL code. A computer program for the interaction of core-melt with concrete. KFK 2890, 1981
6. Hydrogen in water cooled nuclear power reactors. Joint status report of IEAE/CEC. Vienna and Brussels 1990.
7. PRA procedures guide – a guide to the performance of probabilistic risk assessment for nuclear power plants. Vols. 1 and 2 NUREG/CR-2300 (1983)
8. Report of the Special Committee on Source Terms. American Nuclear Society, La Grange Park, Illinois, 1984
9. Report of the American Physical Society of the study group on radionuclide release from severe accidents of nuclear power plants. Draft, February 1985
10. Probabilistic risk assessment (PRA) reference document. Final report. NUREG-1050, September 1984
11. Alsmeyer, H.: BETA experiments in verification of the WECHSL code – experimental results on the melt concrete interaction. Nucl. Eng. Design 103 (1987) 115-125
12. Croff, A.D.: A user's manual for the ORIGEN-2 computer-code. ORNL/TM-7175 (July 1980)
13. Hagen, S. et al.: Results of the CORA experiments on severe fuel damage with and without absorber material. Heat Transfer Conference, Philadelphia, AIChE Symposium 1989, Series No. 262, Vol. 85 p. 135

14. Lorenz, R.A. et al.: Fission product release from highly irradiated fuel heated to 1300–1600 °C in steam. ORNL/NUREG/ TM-346, Dec. 1980
15. Reactor safety study – an assessment of accident risks in US commercial nuclear power plants. WASH-1400 (NUREG-75/014) 1975
16. Deutsche Risikostudie Kernkraftwerke. Eine Untersuchung zu dem durch Störfälle in Kernkraftwerken verursachten Risiko. Köln 1979
17. Deutsche Risikostudie Kernkraftwerke, Fachband 8: Unfallfolgenrechnung und Risikoergebnisse. Köln 1981
18. Randerson, D. (Ed.): Atmospheric science and power production. US Department of Energy, July 1984
19. Pasquill, F.: The estimation of the dispersion of windborne material. Met. Mag. 90 (1961) 33
20. Oberhausen, E.: Die Dosis/Wirkungs-Beziehung bei der Strahlenexposition. GRS-Fachgespräch, Köln, November 1982
21. Fritz-Niggeli, H.: Problematik von Risiskoschätzungen – Beispiel kleiner Strahlendosen. Informationstagung: Sicherheits- und Risikodenken im Zeitalter der Kernenergie, Zürich 1982
22. Lewis, H.W. et al.: Risk assessment group report to the US Nuclear Regulatory Commission, NUREG/CR 0400, 1978
23. Carlson, D.D. et al.: Interim reliability evaluation program procedures guide. NUREG/CR-2728 (SAND-82-1100). Jan. 1983
24. Probabilistic safety analysis procedures guide. Vols. 1 and 2 NUREG/CR–2815-V1-RV-1/XAD and NUREG/CR-2815-V2-RV-1/XAD. Aug. 1985
25. Silberberg, M. et al.: Reassessment of technical bases for estimating source terms. NUREG-0956 (1986)
26. Güldner, W.; Schäfer, H.; Schütz, G.: Entwicklung und Anwendung probabilistischer Methoden für die Sicherheitsbeurteilung in den USA. GRS-A-816 Köln, April 1983
27. Frühauf, H.; Lepie, C.: Overall plant configuration Biblis Nuclear Power Station A and B. Nucl. Eng. Int. 20 (1975) 607-612
28. Kernenergie und Risiko – Fachvortrag – 1. GRS-Fachgespräch München 3/4. November 1977 GRS-10. Köln, März 1978
29. Paschen, H. et al.: Managing risk from nuclear power in the Federal Republic of Germany. The Beijer Institute, Karlsruhe 1982
30. Sizewell B – probabilistic safety study, WCAP 9991 – Rev. 1
31. Kelly, G.N.; Clarke, R.H.: An assessment of the radiological consequences of releases from degraded core accidents for the Sizewell PWR, NRPB-R137. July 1982
32. Gittus, J.: CEGB proof of evidence on: degraded core analysis Sizewell 'B' power station public enquiry. CEGB P 16. November 1982
33. Kelly, G.N. et al.: Degraded core accidents for the Sizewell PWR: A sensitivity analysis of the radiological consequences. NRPB-R 142 October 1982
34. Martz, H.F.; Waller, R.A.: Bayesian reliability analysis. New York 1982
35. Swain, A.D.; Guttman, H.E.: Handbook of human reliability analysis with emphasis on nuclear power plant applications NUREG/CR-1278 (1983)
36. Gittus, J.M.: Degraded core analysis WD-R-610 (S). April 1982
37. Pre-construction safety report for Sizewell B. CEGB (April 1982)
38. Minarick, Z.W.; Kukielka, C.A.: Precursors to potential severe core damage accidents: 1969–1979. A Status Report. NUREG/CR-2497 Vols. 1 and 2 ORNL/NSIC-182, V1 and V2, June 1982
39. Cottrell, W.B.; Minarick, J.W.; Anshin, P.N.; Magen, E.W.; Harris, J.D.: Precursors to potential severe core damage accidents: 1980–1981. A Status Report. NUREG/CR-3591 Vols. 1 and 2, 1984
40. Review of NRC Report: "Precursors to potential severe core damage accidents: 1969–1979. A Status Report" NUREG/CR-2497, Report INPO 82-025, September 1982
41. Hoertner, H.; Kafka, P.; Reichart, G.: The German Precursor Study-methodology and insights. Reliability Engineering & Systems Safety 27, 1990, 53-76
42. Risikoorientierte Analyse zum SNR-300. GRS-51. Köln 1982; Risk-oriented analysis of the SNR-300 – Summary. GRS-56. Köln 1984
43. Nuclear power plant Kalkar (SNR-300). Nuclear Technology 78 (1987) 229-294

44. HTGR accident initiation and progression analysis status report. ERDA-Report, GA-A 13617, I-VII, General Atomic Company, 1975–1976
45. Sicherheitsstudie für Hochtemperaturreaktoren unter deutschen tandortbedingungen, Ergebnisberichte über die Phase Ia. "Kritische Bewertung der AIPA-Studie". Jül-Spez-19, 1978 and Jül-Spez-35, 1979
46. Sicherheitsstudie für HTR-Konzepte unter deutschen Standortbedingungen. Jül-Spez-136, 1981
47. 1160 MWe HTR Demonstrations-Kernkraftwerk. Sicherheitsbericht, Konsortium BBC, HRB 1973/75
48. Sicherheitsgutachten über das 1160 MWe Demonstrationskraftwerk mit Hochtemperaturreaktor. RWTÜV, 1977
49. International Nuclear Fuel Cycle Evaluation (INFCE) IAEA, Vienna 1980
50. Polke, H.; Spindler, H.: "Risikoorienierte Untersuchungen für die Stationen des nuklearen Brennstoffkreislaufes: Ein Statusbericht". GRS-A-873, Köln September 1983
51. Erdmann, R.C. et al.: Status Report on the EPRI fuel cycle accident risk assessment EPRI NP-1128 (1979). Summarized in Nuclear Safety 22 (1981) 300
52. Safety of the nuclear fuel cycle. A state of the art report by a group of experts of the NEA-Committee on the safety of nuclear installations. OECD-Report, Paris (May 1981)
53. The environmental impacts of the production and use of energy: Part II, Nuclear Energy. Energy Report Series ERS-2-79 (1979)
54. Risks associated with nuclear power. A critical review of the literature. Summary and synthesis chapter. National Academy of Sciences, Washington 1979
55. Report to the American Physical Society by the study group on nuclear fuel cycles and waste-management. Rev. Mod. Phys. 50 (January 1978)
56. Sources and effects of ionizing radiation. United Nations Publication No. E. 77.IX.1 (1977)
57. Environmental survey of the uranium fuel cycle. WASH-1248, April 1974. Environmental survey of the reprocessing and waste management portion of the LWR fuel cycle. NUREG-0116 (Suppl. 1 to WASH-1248), 1976
58. Final generic environmental statement on the use of recycle plutonium in mixed oxide fuel in light water cooled reactors. NUREG-0002 (1976)
59. Cohen, S.C.; Dance, K.D.: Scoping assessment of the environmental health risk associated with accidents in the LWR supporting fuel cycle. Teknekron, EPA-68-01-2237, Nov. 1975
60. Draft generic environmental impact statement on uranium milling. NUREG-0511, April 1979
61. Generic environmental impact statement – handling and storage of spent LWR fuel. NUREG-0575, August 1979
62. Storage of spent fuel elements. NEA-Seminar, Madrid 1978
63. Projekt Sicherheitsstudien Entsorgung: Zusammenfassender Abschlußbericht. Berlin, January 1985
64. Benedict, M.; Pigford, T.M.; Levi, H.W.: Nuclear Chemical Engineering. New York 1981
65. Buckner, J.T. et al.: Methodology development for risk assessment of fuel processing. NUREG/CR-1604, Savannah River Laboratory, July 1980
66. Cooperstein, R.; Erdmann, R.C.: Preliminary analysis of a generic fuel reprocessing facility. EPA-520/3-75-003, May 1974
67. Fullwood, R.; Jackson, R.: Actinide partitioning-transmutation program final report: VI. Short-term risk analysis of reprocessing, refabrication and transportation (Summary). ORNL/TM-6986 (March 1980) and Appendices. ORNL/Sub-80/31048/1, Jan. 1980
68. Projektgruppe Andere Entsorgungstechniken: Systemstudie Andere Entsorgungstechniken. Karlsruhe, Dezember 1984
69. Selby, J.M. et al.: Considerations in the assessment of effluents from mixed-oxide fuel fabrication plants. BNWL-1697, June 1973
70. Selby, J.M. et al.: Considerations in the assessment of the consequences of effluents from mixed-oxide fuel fabrication plants. BNWL-1697, Rev. 1, 1975
71. Deutsche Gesellschaft für Wiederaufarbeitung von Kernbrennstoffen: Bericht über das von der Bundesrepublik Deutschland geplante Entsorgungszentrum für ausgediente Brennelemente aus Kernkraftwerken. 2. ergänzte und vollständig überarbeitete Auflage II, September 1977
72. Systemstudie radioaktive Abfälle in der Bundesrepublik Deutschland, Band 1-6 (SRA-1 bis SRA-6). NUKEM im Auftrag des BMFT, Kennzeichen KWA 1214, Hanau 1976

73. Final environmental impact statement. Management of commercially generated radioactive waste. DOE/EIS-0026-F, Oct. 1980
74. Pepping, R.E.: "Risk analysis methodology for spent fuel repositories in bedded salt": Reference repository definition and contributions from handling activities. SAND 81-0219, NUREG/CR-1931, July 1981
75. Claiborne, H.C.; Gera, F.: Potential containment failure mechanisms and their Consequences at a radioactive waste repository in bedded salt in New Mexico. ORNL-TM-4639, Oct. 1974
76. Handling of spent nuclear fuel and final storage of vitrified high-level reprocessing waste. Kärnbränslesäkerhet, Stockholm, Sweden, 1978. Handling and final storage of unreprocessed spent nuclear fuel. Kärnbränslesäkerhet, Stockholm, Sweden, 1978
77. Draft environmental impact statement. Waste isolation pilot plant (WIPP). DOE/EIS-0026-D, April 1979
78. Koplik, C.M. et al.: A status report on risk assessment for nuclear waste disposal. EPRI NP-1197, Oct. 1979
79. Ensminger, D.A. et al.: A review of safety assessment of nuclear waste management. ONWI-126, Nov. 1980
80. Koplik, C.M. et al.: The safety of repositories for highly radioactive wastes. Rev. Mod. Phys. 54 (1982)
81. Kienzler, B. et al.: Untersuchung der Radionuklidausbreitung als Folge eines angenommenen Wasserzutritts zu den Abfällen im Eisenerzbergwerk Konrad. KfK 3410, Okt. 1982
82. Clarke, R.K. et al.: Severities of transportation accidents. Sandia National Labs., SLA-74-0001, July 1976
83. Clarke, R.K. et al.: Severities of transportation accidents involving large packages. Sandia National Labs., SAND 77-0001, 1978
84. Environmental survey of transportation of radioactive materials to and from nuclear power plants. WASH-1238, December 1972; Supplement 1 to WASH-1238. NUREG-75/038, April 1975. Potential releases of cesium from irradiated fuel in a transportation accident, Supplement 2 to WASH-1238. NUREG-0069, July 1976
85. Hodge, C.V.; Jarrett, A.A.: Transportation accidents risks in the nuclear power industry 1975-2000. EPA-520/3-75-023, Nov. 1974
86. Final environmental statement on the transportation of radioactive material by air and other modes. NUREG-0170, December 1977
87. Finley, N.C. et al.: Transport of radionuclides in urban environs: draft environmental assessment. NUREG/CR-0743, SAND 79-0369, Sandia National Labs. (July 1980)
88. Schwachstellen und Risikoabschätzung beim Transport radioaktiver Materialien. Bericht für den BMI, SR 58 (1979)
89. 6th Int. Symp. Packaging and Transportation of Radioactive Materials. PATRAM 80, Berlin, 10.-14. Nov. 1980
90. Rhoads, R.E; Johnson, F.E.: Risks in transporting materials for various energy industries. Nuclear Safety 19 (1978)
91. Rhoads, R.E.; Andrews, W.B.: Transportation risks in the US nuclear fuel cycle. PATRAM 80, Proc. Vol. 1, p. 180, Berlin Nov. 1980
92. Snyder, A.Wm.: A current perspective of the risk significance of steam explosions. Jahrestagung Kerntechnik, Mannheim 1982
93. Karwat, H.: Das Wasserstoffproblem. atomwirtschaft 4 (1983) 186-188
94. Status Report EPRI Hydrogen Program, Feb. 1983
95. Friedrichs, H.G.; Schrödl, E.: Neue Ergebnisse zu Spaltproduktfreisetzungen aus dem Kern und Reaktorgebäude bei Unfällen. GRS-Fachgespräch, Köln 1986
96. Heuser, F.W.: Risikountersuchungen zu Unfällen in Kernkraftwerken. GRS-Fachgespräch, Köln 1986
97. Heuser, F.W.; Hörtner, H.; Kersting, E.: Risikountersuchungen zur Sicherheitsbeurteilung von Kernkraftwerken. Jahrestagung Kerntechnik '87, 2-4 Juni 1987, Karlsruhe
98. Davis, S.M.: Insights gained from probabilistic risk assessments. USNRC, Sept. 1984

4 Risk Studies for Process Plants

Chemical processes involve both physical and chemical hazards. Physical hazards derive from operating conditions which may be extreme, such as very low or very high temperatures and pressures. Chemical hazards are those associated with the materials present in the process, which may be toxic, flammable or explosible, or exhibit several of these properties at the same time. This is complicated further by the fact that some of these properties may vary with changes of process parameters such as temperatures, pressures, or concentrations or that these changes may give rise to unwanted side reactions, as was the case in Seveso [1]. In addition, dangerous properties, if not present under normal process conditions, may evolve upon contact of process media with auxiliary media such as coolants or lubricants. After release, reactions with substances present in the environment, e.g. the humidity of the air, may give rise to dangerous properties.

The procedures in the chemical industry comprise a variety of transport, storage, and transformation processes with differing hazard potentials. Different types of plants are operated; hence, risk analyses have to address the specific circumstances.

Two important risk studies for process plants are the Canvey study [2] and the Rijnmond study [3]. General information about them is given in Table 4.1. These analyses make use of the methods presented in Chap. 2, but have some peculiarities which are described below.

In contrast to nuclear power stations, the structure and employed processes of the process plants investigated in [2] and [3] are simple. Dangerous substances are normally separated from the environment by single barriers, which are already necessary for operational purposes (pipe and tank walls). While dangerous substances in a nuclear power station are concentrated in a few places, the chemical industry uses plants extending over a considerable space. The dangerous substances may be distributed over large areas. This makes retention and protection measures more difficult. In general, the throughput of substances is large and requires transport both inside and outside the plant. This may contribute substantially to total plant risk.

A number of methods are available for identifying sources of hazards and initiating events, some of which were developed specifically for chemical plant analysis. Safety audits, check lists [4], the DOW Fire and Explosion Index [5], Hazard and Operability (HAZOP) Studies [1] and the Failure Mode and Effect

Table 4.1. General information on the risk studies

Characteristic	Canvey	Rijnmond
Originator	Health and Safety Executive at request of the Secretaries for Environment and Employment	Commission for the Safety of the Population at large composed of representatives from government institutions and industry
Analysing team	Health and Safety Executive formed a working group consisting of the Safety, and Reliability Directorate of the UK Atomic Energy Authority, the District Inspector for South Essex, and a Senior Chemical Inspector of Factories	Cremer & Warner Ltd., London
Internal reviewer	None	Battelle-Institut e.V., Frankfurt, SAI and industrial firms involved
Start of the investigation	1976	
Execution time	2 years	2 years
Date of publication	1978	1982 (1980 Report of the steering committee)
Objective	Assessment of the societal risk from the operation of the industrial complex Canvey-Island/Thurrock and its proposed expansion	Judgement on the applicability of the risk analysis methodology to industrial plants and the acquisition of practical experience with its application
Methods	Direct use of statistical data, adaptation of results of fault tree analyses from the literature, subjective judgements, occasional fault tree analyses, models for physical and chemical processes	Direct use of statistical data on undesired events, check lists, Hazard and Operability Studies, fault tree analyses, models for physical and chemical processes
Results	Harm to population in the surroundings of the complex from fires, explosions, and releases of toxic substances, calculation of expected frequencies of occurrence	Harm to employees and population in the surroundings of the complex from fires, explosions, and releases of toxic substances, calculation of expected frequencies of occurrence

Analysis [6] should be mentioned in this context. In addition, these procedures may be used for preparing fault tree analyses (cf. Chap. 2), since many of the steps involved have to be carried out, at least implicitly, in developing fault trees. The use of fault tree analysis for assessing the safety of chemical plants is recent compared to its application to nuclear power stations where it is already state of the art.

4.1   Specific Methods of Analysis

In a safety audit every area of an organization's activity is subjected to systematic critical examination in order to minimize losses. Among others, company policy, personnel training, process properties, plant design and structure, operational procedures, emergency planning, industrial hygiene, and records on previous accidents are considered. The audit serves to identify the strengths and weaknesses and the major areas of hazards of the plant. It is carried out by a team composed of people immediately concerned and professionals, including experts on safety. A formal report and an action plan is worked out after the investigation, whose execution is monitored.

Check lists contain a systematic survey of topics important for plant safety. They represent systematized experience and therefore provide guidance for improving the safety of a plant.

The DOW Fire and Explosion Index is intended to evaluate the risk from fires and explosions caused by the storage, handling, and processing of flammable, combustible, and reactive substances. In particular, the purpose of the procedure, whose quantitative measures are based on historic data, is to

- quantify the expected damage of potential fire and explosion incidents in realistic terms;
- identify equipment which would be likely to contribute to the creation or escalation of an incident;
- communicate the fire and explosion risk potential to management.

To this end the following factors are assessed:

1. Material factor.
2. Factor describing the general process hazards.
3. Factor describing the special process hazards.

The material factor is a measure for possible energy releases of a substance and is calculated from its flammability and reactivity. General process hazards refer, for example, to reaction properties such as endothermic or exothermic. Among the special factors are aspects which may increase hazards like extreme temperatures and pressures or corrosion and erosion. Toxicity is taken into account at this stage as well but only as a limiting factor for emergency response and not as an industrial hygiene or environmental problem. The three factors are combined to a number, the so-called DOW index. This is used to assess the area of exposure and the base maximum probable property damage. In former editions of the index a certain hazard potential was associated with this number, as can be seen in Table 4.2.

Table 4.2. Classification of hazard potentials according to the DOW Fire and Explosion Index

DOW Index	Hazard potential
1– 60	Light
61– 96	Moderate
97–127	Intermediate
128–158	Heavy
>158	Severe

Credit factors which account for safety measures like an emergency shut down system or an emergency power supply are used to modify the index. On the basis of this modified number the maximum credible property damage is obtained from the above mentioned base value, and the maximum probable days of outage of the plant are calculated.

In the Hazard and Operability (HAZOP) Studies which were developed by ICI [7], a multidisciplinary team makes a systematic survey of a process. For every part of the plant possible deviations from the design values, their causes, and their potential consequences are examined. In doing this, a number of guide words like "more", "less", "no", or "inversion" are used and applied to the parameters in question, e.g. more mass flow, less mass flow, no mass flow, flow inversion. These parameters may be process conditions, activities, substances, time, and place. Thus, an overview of possible accident-initiating events and weaknesses of the plant is obtained.

In a Failure Mode and Effect Analysis the important components of the plant are considered, their failures are postulated, and the expected consequences are assessed.

The prerequisite for a successful and efficient application of the aforementioned methods is an accurate knowledge of the plant and its systems, and of the chemical and physical processes involved. The application of systematic methods serves to identify comprehensively the questions to be asked in order to recognize dangerous plant, system, and operating conditions.

As already explained in Chaps. 1 and 2, a risk assessment requires those event sequences to be identified which significantly contribute to risk. The corresponding magnitudes of damage and their expected frequencies of occurrence have to be assessed. The latter are in general derived from observation. Estimates are then either obtained directly from operating experience (e.g. for the frequency of occurrence of pipe leakages), or the undesired event is broken down into sub-events for which operating experience is available. In the latter case the frequency of occurrence is calculated by fault tree analysis. Occasionally accident sequence or event sequence analyses (cf. [8] and Chap. 2) are used for investigating process plants [2, 3]. Nevertheless, frequencies of occurrence must often be established by expert judgement because of a lack of appropriate data. For the event sequences identified as being risk relevant, the materials and energy source terms have to be determined. It must be kept in mind that the search for initiating events may be extremely difficult, as the variety of possible reactions and the material properties may not be precisely known. The materials source term is characterized by the type, quantity, and state of the substance released as well as by the location and duration of the release. For assessing the materials source term the expected leak size has to be calculated and the release process must be analysed, e.g. the discharge of a liquid and subsequent evaporation, the release of a gas or vapour or a two-phase vapour-liquid mixture, or the dispersion of airborne dusts as a consequence of an explosion. If a release directly leads to a fire or an explosion, the quantity of energy released as heat or pressure must be determined.

In studies [2] and [3] only damage to people, but not to property is assessed. In [2] only the effects on the population are considered. Study [3] also provides information on occupational risks. Just early fatalities are taken into account. Long-term and genetic effects which may be expected after a release of a toxic substance are not considered, because appropriate data on the relationship between cause and consequence are not known. Both studies are treated in detail below.

4.2 Canvey [2]

4.2.1 Originator and Objective

The risk study on the industrial complex in the Canvey Island/Thurrock area was carried out between 1976 and 1978 by the Health and Safety Executive of Great Britain at the request of the Secretaries for Environment and Employment. The motive was a public inquiry into the question whether the planning permission to build an oil refinery in the area, given in 1973, should be revoked.

The object of the study was to assess the risk to people resulting from operation of the existing installations and to determine the consequences of the proposed expansion.

4.2.2   Overview of the Industrial Complex and Its Hazard Potentials

Canvey Island is located on the northern bank of the Thames downstream from London. It is inhabited by about 33 000 people and harbours seven important factories with approximately 3200 employees within an area of about $50 \, \text{km}^2$.

The factories mainly store, transform and distribute liquid gas oil, and oil products. Approximately 100 000 tonnes of liquefied natural gas (LNG), 1 880 000 tonnes of oil products, 3 500 000 tonnes of hydrocarbons, 14 000 tonnes of liquefied ammonia, and 500 tonnes of liquefied petroleum gas are stored.

Since some of the substances are explosible, flammable or toxic, their release constitutes a potential hazard to the population. Releases may occur during the transport of the substances by ships, pipelines, road and rail tankers, or their storage, which in some cases is at extremely low temperatures or at pressures between 2 and 8 bar, or during transformation processes.

4.2.3   Investigated Types of Damage

4.2.3.1   Explosions

Gases

Explosions of gas or flammable vapour occur with relative frequency. If they are confined, pressure increases to seven times the initial value are to be expected. Since the open design of the plants under investigation prevents the accumulation of gases, this type of explosion is unlikely to occur. However, there exists the possibility of a vapour cloud explosion. In this context events with releases of approximately 1000 tonnes were considered.

Ammonium Nitrate

In the course of the study it was impossible to determine whether the ammonium nitrate solution present in the plant under investigation can explode or not. Therefore, it was conservatively assumed that explosions are possible.

4.2.3.2   Release of Toxic Substances

No significant way was found by which toxic material could be transmitted to the public via the water supply. After releases of ammonia and hydrogen fluoride (HF) to the atmosphere poisoning is possible. The corresponding model calculations are based on releases of 1000 tonnes of ammonia and 100 tonnes of hydrogen fluoride. According to present knowledge the health effects of hydrogen fluoride are about ten times as severe as those of ammonia; therefore, the health effects of both releases are comparable.

4.2.3.3 Fires

Apart from cases in which a flammable vapour cloud is formed, the public hazard from fires whithin the industrial complex was in general assumed to be low. However, cases in which a fire rages out of control in one installation and spreads to a neighbouring one were considered.

As a consequence of a fire there exists the possibility of release from the lique-fied petroleum gas tanks due to damage caused by radiative heating. The cloud which would then be formed might harm the population in adverse weather con-ditions. Large pool fires of hydrocarbons after spills of about 100 000 tonnes might also affect the population.

4.2.4 Damage-Causing Events and Their Quantification

The question whether and under what circumstances explosions, releases of toxic substances or fires may occur was systematically analysed for all plants. The same problem was examined for transport processes. The following events, which either initiate a release or cause it together with other failures, are mainly to be considered in storage and transformation processes:

– pipe and tank rupture, either spontaneous or due to fatigue;
– rupture of pump casings;
– overpressure in tanks as a consequence of overfilling;
– loss of process control, e.g. pressure, temperature, or flow.

Also, following fire or explosion, damage to neighbouring installations or plants is possible because of

– pressure waves caused by explosions;
– missiles;
– heating-up of tanks.

Flooding and aircraft crashes are the external events whose impact on the industrial complex was considered. For transportation, ship collisions, traffic accidents, and fires during loading and unloading of ships were also investigated.

The frequencies of occurrence of initiating events and the conditional prob-abilities of subsequent events (e.g. the probability of deflagration of a vapour cloud after release) were primarily determined from statistical experience and engineering judgement. Event sequence and fault tree analyses were virtually not used, although they are considered to be the best method for quantifying complex event sequences.

In order to characterize the numerical results, the following classification scheme was used:

(a) Assessed statistically from historical data.
(b) Based on statistics as far as possible but with some missing figures supplied by judgement.

(c) Estimated by comparison with previous cases for which fault tree analyses have been made.
(d) "Dummy figures" – always likely to be uncertain; a subjective judgement must be made.
(f) Analytically-based figures including fault tree analyses which can be independently arrived at by others.

All numbers are associated with uncertainties; their quantification is – except for categories (a) and (f) – rather difficult. In cases (b) and (c) the uncertainties would become smaller if more time and trouble were taken. In category (d), however, prospects are poor for ever reducing the uncertainty. An example is the spill of a flammable liquid as a result of a ship's collision. Will the spill burn in-situ or be carried away as an unburnt cloud?

4.2.5   Impacts of Accidents

Possible sources of explosions in the investigated industrial complex are located more than 1 km away from housing areas. For this reason only minor consequences are to be expected from explosions occurring directly at the location of release.

However, if an explosive vapour cloud is formed, which drifts towards the housing areas and explodes there, the following consequences are likely:

(a) bodily injuries directly due to overpressure;
(b) injuries due to people or things being displaced by the pressure wave;
(c) injuries due to heat, i.e. within the fireball generated by the explosion;
(d) injuries due to widespread fires triggered by the explosion;
(e) death due to asphyxiation;
(f) injuries due to thermal radiation remote from the fireball.

Protection from some of these effects, e.g. thermal radiation, is possible by seeking shelter in buildings.

Based on the population density in the area under consideration (average: 4000 persons km^{-2}), the conditional probabilities of health effects caused by an offsite vapour cloud explosion are estimated, as indicated in Table 4.3. The number of fatalities is considered to be half the number of casualties, i.e. serious injuries or worse. In judging these figures it has to be borne in mind that in order to arrive at the expected frequency of a vapour cloud explosion in an inhabited

Table 4.3. Conditional probabilities for the number of casualties after a vapour cloud explosion

Number of casualties	> 0	1500	3000	4500
Conditional probability	1	0.64	0.35	0.14

area the probabilities still have to be multiplied by the frequency of occurrence of the release of an explosive vapour and the conditional probabilities for the formation of a cloud, its motion in the direction of a housing area, and its ignition in that area.

Casualties may also be caused by emissions of toxic gases. In order to assess them, dispersion calculations both for airborne gases and gases which are denser than air are required. The health effects to be expected depend on both the duration of the exposure and the concentration of the toxic gas.

To assess possible health effects resulting from releases of ammonia, it was assumed that the pressurized storage tank bursts and that a cloud of 1000 tonnes of anhydrous ammonia is formed, which initially is composed of 20% vapour and 80% liquid. The cloud spreads in accordance with the topographic situation at the location of release and the weather conditions at that time. For average weather conditions in the area and a windspeed of 6 m/s dangerous ammonia concentrations are to be expected in an elliptically shaped zone with half axes of 2.5 and 3 km oriented in the direction of the wind. Accounting for the population distribution and averaging over the wind directions to be expected, the conditional probabilities of Table 4.4 are obtained. They represent the number of persons exposed to hazards from ammonia. It has to be borne in mind that damage to health or deaths are not necessarily to be expected in every member of the affected population, since the possibility of mitigating measures like staying indoors or medical treatment has not been taken into account.

Table 4.4. Conditional probabilities for exposure to intoxication risk in the Canvey Island/Thurrock Area following a release of 1000 tonnes of ammonia

No. of people at risk	Conditional probability
<100	0.59
100– 1 000	0.11
1 000– 2 000	0.02
2 000– 5 000	0.03
5 000–10 000	0.07
10 000–20 000	0.02
20 000–30 000	0.14

In order to calculate expected frequencies, the figures of Table 4.4 still have to be multiplied by the expected frequency of the burst of the pressurized storage tank, which is assumed to be $10^{-5}\,a^{-1}$. When judging these results it has to be kept in mind that significant uncertainties are associated with them.

The release of hydrogen fluoride was not investigated in detail. As previously mentioned, it was supposed that on release of 100 tonnes of this substance similar effects as in case of a release of 1000 tonnes of ammonia are to be expected.

It was assumed that fires normally do not endanger the public, since they are mostly confined to a small area, e.g. the bunds below the storage tanks, which would contain the spilled flammable substances. Nevertheless, the final results of Table 4.5 include some cases in which flammable liquids are released in large quantities following an explosion close to their place of storage. Casualties among the population have then to be expected.

Emergency schemes for the area under consideration are available to deal with a limited release of ammonia. A release of 1000 tonnes, however, would require a wide-scale evacuation. A corresponding plan only exists to deal with flooding which may overtop the sea defences, in which case the pre-warning times are much longer (12 h to 24 h) than in the case of accidental releases. With some scenarios, however, allowance was made for a large-scale evacuation of the population, if a warning could be given two hours in advance. It was then assumed that half the population is evacuated in the first two hours after the warning. During the subsequent two hours another half of the remainder of the population is evacuated, etc.

4.2.6   Results and Conclusions

Thirty eight cases, each of which comprises several accident scenarios were investigated. An expected frequency of occurrence of severe accidents of $3 \cdot 10^{-2}\,a^{-1}$ was obtained. However, not every accident affects the population. Table 4.5 shows the number of casualties and corresponding frequencies to be expected. Results for the present state of the industrial complex, and also the reduction of the societal risk after introducing the proposed improvements, are shown. The maximum individual risks are indicated in Table 4.6. It should be noted that the figures refer to casualties and that the number of fatalities would roughly be smaller by a factor of 2.

Table 4.5. Expected frequencies for societal risk from casualties

No. of casualties >	Existing installations		Existing installations plus proposed developments	
	Expected frequency in $10^{-4}\,a^{-1}$	Expected frequency after improvements in $10^{-4}\,a^{-1}$	Expected frequency in $10^{-4}\,a^{-1}$	Expected frequency after improvements in $10^{-4}\,a^{-1}$
10	31.4	8.6	47.5	10.8
1 500	17	4.2	29.1	5.7
3 000	10.8	2.9	17.7	3.8
4 500	6.1	2	9.3	2.6
6 000	3	1	4.4	1.3
12 000	1.7	0.6	2.7	0.8
18 000	1	0.3	1.8	0.5

Table 4.6. Maximum individual risk from casualties

	Expected frequency in 10^{-4} a^{-1}	
	Existing installations	Existing installations plus proposed developments
Before improvements	13	26.3
After improvements	6.1	7.7

The authors emphasize, that owing to numerous uncertainties the results may only be considered as gross estimates.

It is evident that the proposed modifications to the existing plants reduce the societal and individual risks to such an extent that even after the planned extension the risk associated with the previous state is not reached. Figure 4.1 shows the societal risk for the entire industrial complex in terms of casualties before and after the proposed improvements.

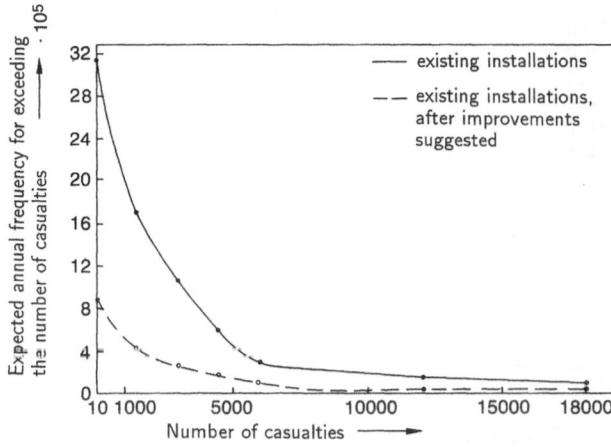

Fig. 4.1. Societal risk in terms of casualties for the Canvey/Thurrock area

4.2.7 Canvey – A Second Report

Three years after the publication of [2] a reassessment of the risks from the various industrial operations in the Canvey/Thurrock area was published [9]. It reflects the situation in 1981 after a number of improvements were implemented and accounts for several others still to be introduced. The main factors which have contributed to the reassessed values are:

– physical improvements, e.g. to liquefied natural gas ships;
– changes in operation, e.g. cessation of ammonia storage at one of the firms;

- detailed studies by firms which have made a properly argued and documented case for reducing a risk previously assessed on a provisional basis, e.g. the risk from a limited spill of ammonia at one of the firms;
- changes in assessment techniques, e.g. improved understanding of the dispersion of heavy gases;
- correction of occasional errors in [2];
- additional work or changes in operations in the future for which there is a firm commitment, e.g. decommissioning of in-ground LNG tanks.

Table 4.7 gives the reassessed figures for the societal risk applicable to the industrial complex with extensions and after the implementation of the improvements.

The figures are somewhat lower than those of Table 4.5.

Table 4.7. Expected frequencies for societal risk in terms of casualties after reassessment

Number of casualties >	Expected frequency in 10^{-4} a^{-1}
10	10.0
1 500	2.0
3 000	0.8
4 500	0.4
6 000	0.2
12 000	0.1
18 000	0.1

4.3 Rijnmond [3]

4.3.1 Originator and Objectives

The risk study for the Rijnmond area was carried out at the request of the Commission for the Safety of the Population of the Netherlands (COVO). The commission is composed of representatives of government authorities and industry, and advises the executive board of the Rijnmond Authority on industrial safety regarding the population. A project committee composed of representatives of the authorities and the industrial companies involved was formed to direct the study. Whenever necessary, experts on certain subjects and representatives of the firms performing the study (Cremer & Warner, London, and Battelle, Frankfurt) were consulted. It took two years to complete the investigation. Its results were published in 1982.

In the study the hazards to the employees and the population from the operation of six selected installations in the Rijnmond area were analysed. The installations were chosen in such a way that releases of toxic and flammable substances, and their storage at low temperature and under pressure had to be treated. The investigation is considered to be a pilot study; its objective was to evaluate the feasability of risk analyses for industrial plants and to gain experience in their practical application.

Particular objectives were to assess the quality of accident investigations and of probability evaluations, given the state of risk analyses, and to identify existing problems and gaps in the field of risk analyses for chemical plants.

The results of the investigations were to form a basis for deciding about the role of risk analysis in formulating safety policies. Furthermore, a way of presenting the results with regard to their use as a decision-making tool was to be found. The effort in terms of time and money required to assess industrial risks with sufficient accuracy was to be determined.

4.3.2 Overview of the Industrial Complex and Its Hazard Potentials

The Rijnmond area is the delta of the Rhine stretching from Rotterdam to the North Sea. It is about 40 km long and 15 km wide with 1 000 000 people living in the area. The largest harbour in the world is situated there with its high density of chemical and petrochemical industrial plants, some of which are close to residential areas.

The six study objects selected comprise the storage of toxic and flammable substances and part of a plant for a chemical separation process.

(1) Acrylonitrile (ACN) Storage

A tank of 3700 m^3 capacity and the corresponding transfer system of pipelines and pumps for delivering ACN to and from the tank were analysed. Road and rail tankers, and barges were not included in the investigation. Therefore, the possibility of overfilling them, which had contributed significantly to accidents in the plant, was not taken into account. The storage is at ambient temperature and practically at atmospheric pressure.

The tank is equipped with a system to extinguish tank fires and to cool its walls. The installation is mainly operated by local manual control equipment. Only few central measurement indicators and controls are installed.

(2) Ammonia Storage

A spherical tank for ammonia storage which belongs to a plant for producing different chemical raw materials and fertilizers was analysed together with the associated pipelines, pumps, and ancillary equipment for ammonia transfer to and from the storage sphere. Ammonia production as well as the road and railtankers and tank ships related to the system were not considered. The sphere stores liquefied ammonia under pressure (up to 12 bar) and at ambient temperature. The tank has a volume of $1000\,m^3$ and an average inventory of 250 000 kg, which corresponds to 40% of its total capacity. An emergency shut-down system is installed, which is activated in the event of certain hazardous occurrences (such as high level in the sphere). Normal operation requires the co-ordination of local manual intervention and remotely controlled actions.

(3) Chlorine Storage

The storage of liquid chlorine at ambient temperature under a pressure of about 6.5 bar is part of a bigger chemical plant. The storage facility, consisting of five tanks of $90\,m^3$ capacity each, which corresponds to 100 t of liquid chlorine, and the boil-off compressor and pipelines were analysed. Of the five tanks one is always kept empty to act as a relief tank after the breach of the corresponding rupture disks in case of overpressure. The valves are positioned in such a way that the empty tank can always take up the entire contents of any one of the other tanks. The effective capacity of the pressurized storage system is therefore 400 tonnes of chlorine. There are several systems for pressure and level control, some with indicators in the control room. The alarm pressure gauge is set at 11 bar. The throughput of the facility is about 300 tonnes of chlorine a day.

(4) Liquefied Natural Gas (LNG) Storage

The fourth object of investigation is the storage of liquefied natural gas. It forms part of a plant which provides gas to the mains during periods of heavy demand in the winter. On demand the gas is fed into the mains after adding certain quantities of nitrogen. Two LNG tanks were investigated together with their auxiliary equipment for operation, loading, and unloading.

Each tank has a capacity of $57\,000\,m^3$ LNG. Since the LNG is stored at a temperature of $-162\,°C$, the tanks are double walled. Between the inner and the outer tank there is an insulating layer which is more than 1 m thick. The entire tank stands in an open cylindrical concrete containment, which is designed to withstand the thermal shocks arising from the release of the whole contents of the tank.

Pressure increases due to heat leakage into the tank are compensated by drawing off the boiled-off vapour and delivering it to the gas mains. Severe excursions of pressure are dealt with by a staggered system of vent lines, relief valves, and emergency valves against overpressurization and by gas injection and vacuum breakers in case of underpressure. In addition, the possibility of an emergency shut-down of either individual parts of the plant or the plant as a whole is provided for. Critical areas of the plant are monitored by a network of gas detectors. When the set points are exceeded, heaters are turned off and the air inlets of the compressor house and of the control room are closed.

(5) Propylene Storage

The storage of propylene analysed in the study is part of a factory for the production of propylene oxide. Two spherical storage tanks were investigated and the corresponding pipelines and jetty facilities for unloading ships and railcars with their vaporizers and lines to storage, as well as the supply line to the propylene oxide plant. The two spherical tanks with a combined capacity of 1200 tonnes function at a maximum pressure of 14 bar at ambient temperature. The storage system is operated almost entirely by local manual control of valves and other components. The system parameters are only indicated locally.

(6) Hydrodesulphurizer

The diethanol amine (DEA) regenerator is part of a hydrodesulphurizer for removing sulphur from gasoil. In this process sour gas containing hydrogen sulphide H_2S is formed. This is removed in order to produce a sulphur-free fuel gas. The removal process is regenerative and uses DEA.

The regenerator is operated at 60% of its design load and at approximately 92 °C and 0.6 bar overpressure at the top of the column. The pressure in the system is controlled by control valves; the H_2S is either recovered or eliminated by a flare system, which also allows excessive pressure to be relieved via pressure relief valves on top of the column. Containers and pipes with unregenerated DEA may be discharged to a sewage system.

The H_2S tightness of the plant is monitored by detectors at certain locations such as gland seals on pumps. In case of leakage a sound alarm is activated. However, the location of the release has to be found by an operator who is then sent to the plant. If a major release of H_2S were to occur, the flow of H_2S can quickly be reduced from the control room by shutting off the steam supply to the process and the pumps.

4.3.3   Types of Damage Investigated

The analysis of accidental damage in process plants generally requires modelling of the effects of

- explosions,
- fires, and
- releases of toxic substances

in order to assess their impact on employees and the public.

Knowledge about explosions of vapour clouds is insufficient. For this reason several models were compared and a semi-empirical model, the so-called TNO-Model [10], was chosen. It is believed that this comes closest to the requirements. Nevertheless, doubts remain as to whether the combustion process and the flame propagation speed are correctly modelled. The assessment of damage caused by shock waves only accounts for pressure peaks; the possibility of varying pressure-time-histories and of the formation of missiles was not considered.

Models are given for pool fires, locally confined fires, and flash fires. The damage caused by them is classified according to the heat flux density. Possible burns from the fire ball produced by a boiling liquid expanding vapour explosion (BLEVE) are treated separately.

The inhalation of toxic substances may cause various effects ranging from mild irritations of the respiratory tract to death. The possibility of death depends not only on concentration, but also on the duration of the exposure. The dependence between the two parameters is, in general, non-linear, so that with increasing concentration the times of exposure until death drop rapidly. The correct functional dependence is not known for most chemical substances.

Toxic effects in dependence on concentration and duration of exposure were assessed for the following substances:

- chlorine,
- ammonia,
- hydrogen sulphide,
- acrylonitrile.

4.3.4 Method of Analysis

4.3.4.1 Initiating and Undesired Events

In all plants under consideration dangerous substances are contained during normal operation. A hazard to the environment will only arise in the case of failure of the containment. In order to determine the initiating events the distribution of the dangerous substances in the plant was examined.

To identify the failure modes to be investigated, possible failures of barriers were analysed using checklists and in some cases Hazard and Operability Studies (cf. Sect. 4.1). Failure modes leading to similar release scenarios were combined. Different operational states and the possibility of protective measures by the safety systems were accounted for. Thus the initiating events for the plants under consideration were identified. Their numbers and the corresponding undesired events are listed in Table 4.8.

The expected frequencies of the majority of undesired events were estimated directly from statistical data. Only for those parts of the installations whose safety depends on complex control systems was the more laborious method of fault tree analysis used in order to deduce the expected frequencies of the undesired events from the expected frequencies of the initiating events and the probabilities of component failures and human errors. Fires, earthquakes, vehicle impact, subsidence, flooding, and mechanical impacts are the external events treated.

Table 4.8. Initiating and undesired events for the analysed plants

Plant	Number of initiating events	Undesired event(s)
Acrylonitrile	28	Toxic effects, fires, explosions
Ammonia	17	Toxic effects
Chlorine	40	Toxic effects
LNG	6	Fires
Propylene	31	Fires, explosions
Hydrodesulphurizer	6	Toxic effects

4.3.4.2 Data for Quantifying Undesired Events

The expected frequencies of undesired events were either taken from existing statistics or assessed with the help of fault tree analyses. For this purpose the expected frequencies of initiating events and the probabilities of component failures and human errors were required. These were taken from three sources:

- Data collections in the literature;
- Data supplied by the plant owners (especially on different operational states);
- Estimates.

Uncertainty of the input data was described by log-normal distributions. If the failure behaviour was expressed in terms of an unavailability, this was calculated according to the asymptotic relation for components under repair:

$$\text{unavailability} = \frac{\text{downtime}}{\text{downtime} + \text{mean time to failure}} \quad (4.1)$$

The duration of the downtime depends on factors such as the time until the discovery of the failure, the time required for repair, and the availability of spare parts and maintenance personnel. The corresponding figures were provided by the plant owner or, if not available, estimated from data in the literature. As mentioned before, the mean time to failure is the reciprocal value of the failure rate. The failure rates given in the literature were considered as base data which were adapted to the special load conditions in the investigated plants using environmental factors.

The following areas were treated:

- pumps,
- pipework,
- hoses,
- loading arms,
- valves,
- measuring instruments,
- control and transmitter equipment,
- general electrical equipment,

- tanks,
- human errors,
- external events.

Data on the reliability of components and the failure modes to be considered were taken from compilations for non-nuclear and nuclear technology. Data for human errors during normal operation, maintenance, and emergency situations were taken from the literature. They formed the basis of case-by-case estimates of the values to be used.

The expected frequencies of occurrence of external events and their distributions are also given. The figures, however, are characterized as rough estimates.

No data on possible common cause failures is given, since these are not taken into account in the analysis.

4.3.4.3 Fault Tree Analyses

If the initiating event causes the undesired event directly (e.g. spontaneous pipe rupture), the expected frequency of occurrence of the latter was taken from the corresponding statistics. In other cases fault tree analysis was used. In view of the considerable effort involved this was only deemed appropriate when the safe containment of dangerous substances depends on complex control systems.

The fault trees were evaluated by determining their minimal cut sets and the corresponding structure functions (cf. Sect. 2.1.4.2). Uncertainties of basic event data (initiating events, component failures, human error, external events) were accounted for by taking random samples from their corresponding probability distributions. Thus, in addition to the mean value, 5% and 95% centiles of the frequency distribution of the undesired event were calculated. The intervals thus obtained reflect the uncertainty in quantifying the basic events. The computer program used can only treat fault trees with less than 500 minimal cut sets, so that in some cases the fault trees had to be divided into smaller units, which were evaluated separately.

The following possible sources of error in the fault tree analysis are mentioned:

- incorrect estimation of probabilities and frequencies of basic events;
- the hypothesis of independent basic events, even if unjustified;
- incomplete fault trees.

4.3.4.4 Determination of Source Terms

In order to determine the quantities of released substances and the time histories of their releases, the discharge processes and the evaporation of liquids from pools were modelled. Care was taken to select models based as far as possible on first principles with little dependence on empirical correlations.

In order to determine the discharge rates the failure cases were classified. Depending on the location of release and the type of containment, six different model calculations were used. They are characterized as state-of-the-art engineering calculations, whose accuracy is sufficient in view of the uncertainties of the input data. In the case of two-phase critical flow discharge rigorous model calculations are not possible. Instead empirical correlations were used.

In order to determine the vaporization rates of released liquid dangerous substances, the spread of pools of liquids on different types of ground and water surfaces was investigated for each type of plant.

The vaporization rates were assessed as a function of the material properties and of the different mechanisms of heat transfer. The models used for that purpose are based on improved versions of procedures reported in the literature, particularly in the "yellow book" of TNO [10].

If no generally accepted theory was available (e.g. for describing the evaporation behaviour of liquefied gases), or if especially unfavourable boundary conditions had to be treated (e.g. jets directed towards the ground), the corresponding parameters were estimated conservatively.

4.3.4.5 Dispersion Calculations

Given the different types of release and the differing physical and chemical properties of the materials involved, various types of clouds may be formed after a release. Three simplified cases were treated in the dispersion calculations:

– Instantaneous release with formation of a cloud of cylindrical shape (this was used, for example, to treat dispersion after a catastrophic rupture of the pressurized propylene sphere).
– A steady release of finite duration in the form of a jet which could be angled in any direction within the vertical plane through the wind direction (this was used, for example, after the double ended break of a propylene pipeline).
– A steady release of finite duration having negligible initial momentum (e.g. for treating the boil-off of vapour from an LNG spill).

The density of the released substance is important in the dispersion of the cloud. For LNG vapour, chlorine, propylene, and acrylonitrile the cloud will always be significantly denser than air after release. In the case of ammonia, due to the low molecular weight, a buoyant cloud may be formed. Because of aerosol formation and temperature reduction due to the evaporation of liquid, however, the cloud will usually be denser than air just after release. The dispersion of H_2S may be calculated using the density of air. The effect of the initial momentum on the plume behaviour will vary considerably depending on the form of release, the orientation of the jet, the possible presence of obstacles, and other factors. In general, it may be assumed that a cloud will eventually subside towards the ground and that its motion will become gravity-dominated rather than momentum-dominated. In the long run the effect of atmospheric turbulence will become dominant and the cloud will be dispersed.

In order to treat these different release scenarios, two different dispersion models were used, which describe the dispersion of jet type releases with initial momentum and the gravity spreading of dense clouds with ground contact.

Both models take into account the density of the cloud and that of the sur-rounding air. In one of the models a sudden transition from gravity-dominated to turbulence-dominated dispersion is assumed; in the other this is achieved through the inclusion of atmospheric turbulence effects throughout the cal-culation. In the case of neutral density the Gaussian dispersion model was adopted, accounting for different classes of atmospheric stabilty and a surface roughness which reflects local topography.

The meterological data required for the dispersion calculations were taken from records of the meteorological stations at Zestienhoven (for five plants) and Hoek van Holland (for one plant). From the recorded five wind speed ranges and six stability classes, six (eight for the Hoek van Holland data) representative weather categories were formed in order to reduce the volume of calculations. Since the models used cannot treat calm weather or varying wind directions; these cases, which comprise about 10% of all weather situations, were included in the lowest wind speed range. Possible mitigating effects such as the impact of rainfall were not accounted for.

4.3.5   Impacts of Accidents and Counter-measures

4.3.5.1   Explosions

As mentioned above, only the effects of blast waves following explosions were treated. The correlation between the magnitude of damage and the peak over-pressure, as given in Table 4.9, was then used.

Table 4.9. Assignment of the magnitudes of damage to the peak overpressure in a blast wave

Magnitude of damage	Blast overpressure Δp in bar
Major structural damage (assumed fatal to people inside buildings or other structures)	0.3
Repairable damage. Pressure vessels remain intact; light structures collapse	0.1
Windows break, possibly causing some injuries	0.03

4.3.5.2   Fires

At the location of release two types of fires were distinguished, a continuous pool or jet fire and the immediate ignition of released pressurized fuel (BLEVE) leading to a fire ball. The thermal radiation from pool and jet fires was calcu-lated by adapting the theoretical models to the empirical data available from experiments and observations. The same applies for the calculation of the effects of fire balls, where the semi-empirical estimation of their energetic effects was preferred to applying dynamic models. Stationary as well as transient fires are possible, the latter, in particular, after a vapour cloud explosion. For stationary fires the correlations between the magnitude of damage and heat flux density due to radiation from the fire of Table 4.10 were used. In the case of a fire ball caused by a boiling liquid expanding vapour explosion the correlations for stationary fires of Table 4.10 may not be used because of its short duration (5–30 s). Table 4.11 contains the relationship between the type and magnitude of damage in dependence on the energy flux applicable in this case.

Table 4.10. Assignment of the type and magnitude of damage to the heat flux density levels for steady state fires

Type and magnitude of damage	Heat flux in kW/m^2
Sufficient to cause damage to process equipment	37.5
Minimum energy required to ignite wood at infinitely long exposures (non-piloted)	25
Minimum energy required for piloted ignition of wood, melting plastic tubing etc.	12.5
Sufficient to cause pain to personnel, if unable to reach cover within 20 s; however, blistering of skin (first degree burns) is likely	4.5
Will cause no discomfort even with long lasting exposures	1.6

Table 4.11. Assignment of the type and magnitude of damage to the energy flux density for vapour cloud explosions (fireballs)

Type and magnitude of damage	Energy flux density in kJ/m^2
Third degree burns	>375
Second degree burns	>250
First degree burns	>125
Threshold of pain, no reddening or blistering of skin caused below this value	>65

4.3.5.3 Toxic Vapours and Gases

The consequences of inhaling toxic gases range from mild irritations of the respiratory tract to death. The effect depends on the duration of exposure and on the concentration of the toxic substance. The relationships between exposure and consequences for the different materials involved are shown in Tables 4.12 to 4.14.

Little is known of the toxic effect of acrylonitrile on humans. For this reason results from experiments with animals were used. A lethal dose value LD50 (i.e. a toxic load by which 50% of the exposed population is killed) of $1.5 \cdot 10^4$ ppm $\times$ min was chosen (product of concentration and duration of exposure in minutes).

The effect of toxic gases may be reduced by going indoors and remaining there while the gas cloud passes. In a modern building the air is exchanged once per hour (two to three or more if doors or windows are ajar). The calculations were based on two air changes per hour and on the assumption that 99% of the population are indoors or seek refuge in buildings shortly after the release. To

Table 4.12. Toxic effects of chlorine in dependence on concentration and duration of exposure

Effect	Duration of exposure	Concentration in ppm
Odour detectable by most people	Any	1.0
Negligible effects – mild irritation	Any	<3.0
Serious distress – strong irritation	Any	5 ... 20
Lethal	A few breaths	1000
	< 15 mins	>75
	30 ... 60 min	40 ... 60
	60 ... 90 min	35 ... 51

Table 4.13. Toxic effects of ammonia in dependence on concentration and duration of exposure

Effect	Duration of exposure	Concentration in ppm
Odour detectable by most people	Any	25
Negligible effects – mild irritation	Any	<100
Serious distress – strong irritation	Any	300 ... 500
Lethal	A few breaths	>5000
	<15 min	2000 ... 5000
	30 ... 60 min	>1700

Table 4.14. Toxic effects of hydrogen sulphide in dependence on concentration and duration of exposure

Effect	Duration of exposure	Concentration in ppm
Odour detectable by most people	Any	0.1 ... 0.4
Safe for 8 hours of exposure	8 h	10
Maximum which can be inhaled without serious consequences	60 min	200
Lethal	Rapidly	>900
	<30 min	600 ... 800

estimate the number of workers who are outdoors or indoors at the moment of release the characteristics of their jobs were used.

4.3.6 Results and Conclusions

Table 4.15 shows the overall levels of risk to employees and the population. When judging the results it must be borne in mind that the figures only apply to those parts of the plants which were analysed. They are not representative of the risk of the entire production processes in question. Evidently, the risks of the different study objects differ widely.

Table 4.15. Total risk from the study objects

Study object	Number of fatalities per year		Employees' individual risk of death per calendar year
	Employees	Population	
Acrylonitrile	$2.1 \cdot 10^{-3}$	$7.9 \cdot 10^{-6}$	$6.6 \cdot 10^{-6}$
Ammonia	$2.1 \cdot 10^{-3}$	$2.0 \cdot 10^{-4}$	$2.0 \cdot 10^{-6}$
Chlorine	$1.1 \cdot 10^{-2}$	$3.6 \cdot 10^{-3}$	$5.1 \cdot 10^{-5}$
LNG	$1.5 \cdot 10^{-7}$	$6.8 \cdot 10^{-10}$	$5.7 \cdot 10^{-9}$
Propylene	$1.1 \cdot 10^{-4}$	$3.7 \cdot 10^{-5}$	$7.7 \cdot 10^{-7}$
Hydrodesulphurizer	$1.0 \cdot 10^{-6}$	0	$2.1 \cdot 10^{-9}$

The risk from the hydrodesulphurizer is low, since the hazard potential of the substances involved is low and the plant is well designed.

The plant for liquid natural gas has a low population risk due to its remote location away from housing areas and the concrete containment enclosing the gas tank.

Risk to the population from the acrylonitrile plant is very low, since the hazards resulting from its operation are of short range. On the other hand, the employee risks are comparatively high, since accidents are relatively frequent.

For ammonia, chlorine, and propylene the risks are high compared to those of the other plants. The main reasons for this are the large hazard potential of the stored substances, the quantities involved, their location close to residential areas, and the characteristic behaviour of pressurized liquefied gases following release.

The results (with the exception of Table 4.15) are presented in the following way:

- individual risks: by a plot of lines of constant risk drawn on a map covering the geographical area in which damage could occur, i.e. "iso-risk" lines (Figs. 4.2 and 4.3);
- population risks: by means of complementary cumulative frequency distributions (Figs. 4.4 to 4.9).

In conclusion it is stated that the methods employed are appropriate in principle and effective in practice. They may, however, be too detailed and laborious for widespread use. On the other hand, they could readily be simplified, for example by reducing the number of weather cases, wind directions, and failure scenarios.

4.3.7 Uncertainties

As explained in Chap. 2, uncertainties are associated with risk assessment. They concern the calculation of the frequencies of occurrence of the undesired events and the estimation of the corresponding magnitudes of damage.

In the Rijnmond study the calculation of the frequencies of occurrence of damage is the main contributor to uncertainties in the final result. The influence of any uncertainties when modelling physical and chemical processes inside the plant is regarded as smaller.

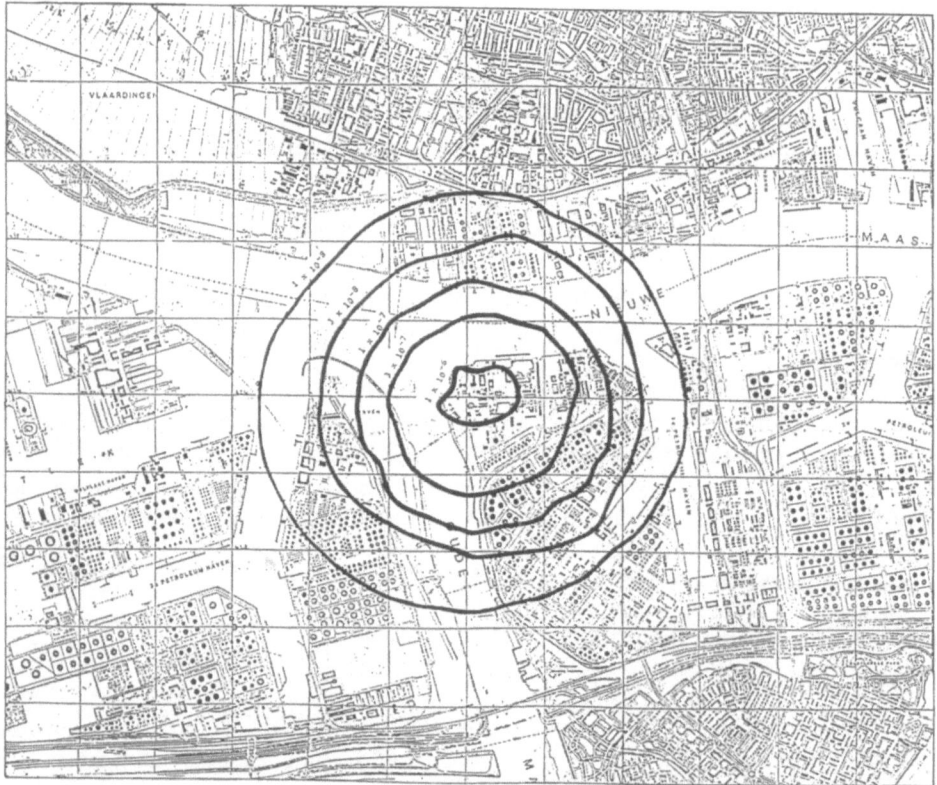

Fig. 4.2. Iso-risk lines for indoor exposure after releases from the ammonia storage (all release cases). Copyright © 1982 by D. Reidel Publishing Company, Dordrecht, Holland

Important sources of error stem from treatment of the effects of toxic substances and of the atmospheric dispersion of vapours and gases. Further uncertainties are associated with assumptions concerning evacuation measures. The uncertainties of the expected number of fatalities per year are indicated in Table 4.16. The values are subjective estimates; they cannot – as is emphasized – be interpreted as 95% confidence intervals.

Table 4.16. Uncertainties of the results

Study object	Uncertainty
Acrylonitrile	One order of magnitude
Ammonia	About one order of magnitude
Chlorine	Somewhat more than one order of magnitude
LNG	One half to one order of magnitude
Propylene	Above one order of magnitude
Hydrodesulphurizer	Above one order of magnitude

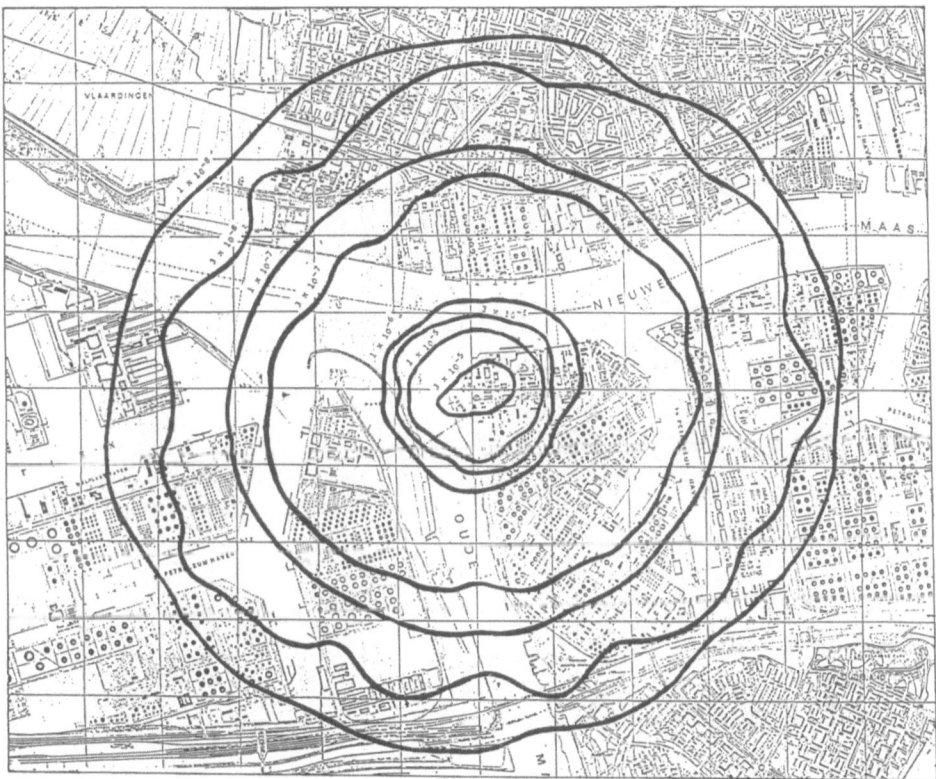

Fig. 4.3. Iso-risk lines for outdoor exposure after releases from the ammonia storage (all release cases). Copyright © 1982 by D. Reidel Publishing Company, Dordrecht, Holland

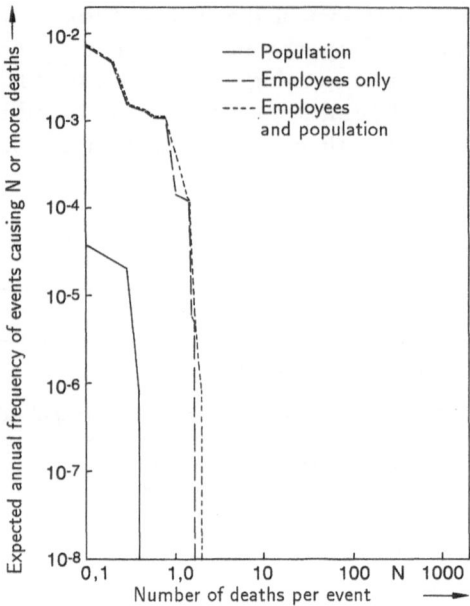

Fig. 4.4. Complementary cumulative frequency distribution for deaths from storage of acrylonitrile

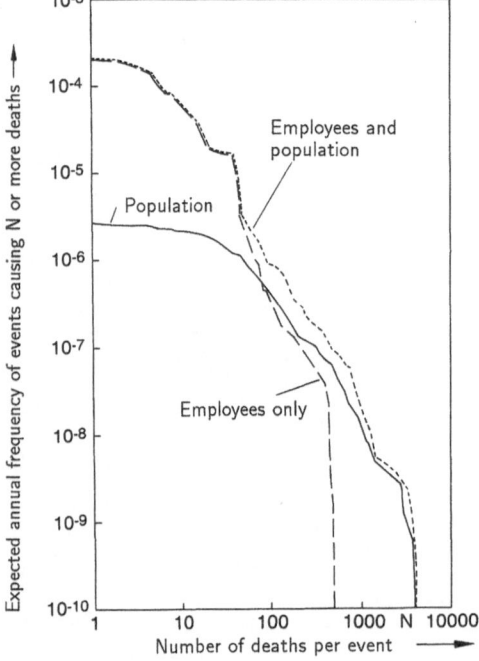

Fig. 4.5. Complementary cumulative frequency distribution for deaths from storage of ammonia

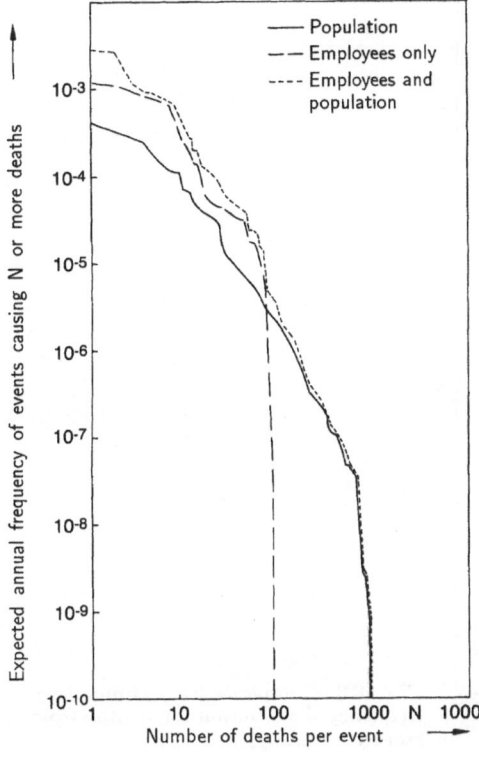

Fig. 4.6. Complementary cumulative frequency distribution for deaths from storage of chlorine

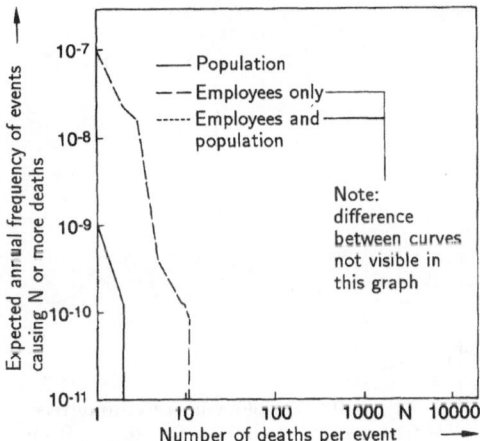

Fig. 4.7. Complementary cumulative frequency distribution for deaths from storage of LNG

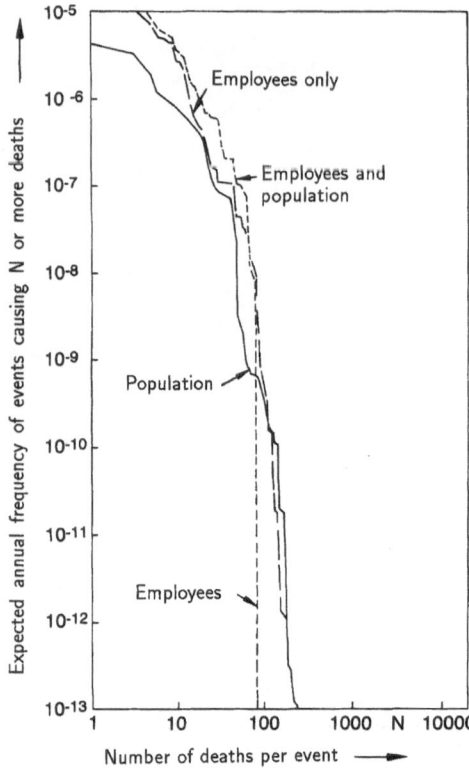

Fig. 4.8. Complementary cumulative frequency distribution for deaths from storage of propylene

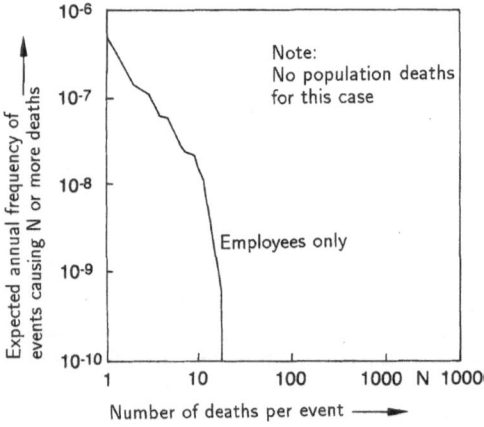

Fig. 4.9. Complementary cumulative frequency distribution for deaths from operation of the hydrodesulphurizer

The comments from industry [3] put some of the figures in doubt. The risk numbers are believed to be too high and the uncertainty ranges too narrow.

4.4 Comparative Remarks

4.4.1 Canvey-Rijnmond

As already shown, the objectives of both studies are different. Whilst in the Canvey report the population risk from the operation of the entire industrial complex was assessed, the objective of the Rijmond study was to judge to what extent risk analysis methods can be applied to the investigation of chemical plants. Accordingly, the emphasis is different. In [2] detailed analyses of individual plants were ignored in favour of a global assessment on the basis of available statistics or estimates. Emphasis was placed on consequential damage which may occur in one plant following incidents in another plant on the industrial complex.

In [3], on the other hand, selected study objects were investigated in considerable detail using fault tree analyses. Hence, the results are not an evaluation of the entire complex, since the study objects are only part of the plants. The impacts of events on other installations within the industrial complex were not included.

The differences mentioned might explain why the results of both investigations differ by several orders of magnitude. Any certainty could only be reached if the procedure of one study was applied to the industrial complex of the other.

The method of [2] could be described as "macroscopic". As already mentioned, it ignores details in order to appraise the entire complex with practicable effort. For this reason the proposals for improvement are global and do not address plant specific details. The procedure in [3], however, focuses on details of plant design and reveals weaknesses. Both procedures may be considered as largely complementary. Analyses of the type carried out in [3] might serve for the "estimation by comparison with previous cases, for which fault tree analyses have been made", as used in [2] (cf. Sect. 4.2.4). Ideally a combination of both procedures should be employed, reserving fault tree analyses for technically complex and hazardous installations and analysing other plants on the basis of statistical data and, if unavoidable, by engineering judgement. In any case, the possible impact of an incident in one plant on other installations within the industrial complex should be accounted for.

Owing to these differences between objectives and methods the results of both studies may not be compared with each other.

4.4.2   Rijnmond – DRS-A

4.4.2.1   Preliminary Remark

Since the procedure in [3] resembles that of the German risk study for nuclear
power plants (DRS-A) [11], the differences and similarities of both studies are
explained below.

4.4.2.2   Objective

While the Rijnmond study was to explore the applicability of probabilistic
methods to risk analysis of industrial plants, the DRS-A could rely on the
comprehensive investigation of the risk of nuclear power plants in [12] and take
over its approach as far as was reasonable (cf. Sect. 3.2.4.1).

4.4.2.3   Object of the Analyses

The technical designs of the chemical plants and the nuclear power plant investigated differ considerably. This is due in part to the processes and in part to
the different safety concepts. Differences chiefly exist in the local distribution
of dangerous substances, the operational concept, and the number of safety
systems. In addition, the mechanisms of damage of the dangerous materials
involved are different.

During the normal operation of a nuclear power plant the dangerous substances are concentrated almost entirely in the reactor core. The importance
attributed to the integrity of the fuel elements reflects the safety concept of
the nuclear power industry to apply damage-preventing measures as closely as
possible to the source. In the process plants under consideration this concept
cannot be applied to the same extent, since the process media themselves are
the dangerous substances and the quantities involved and their throughput
render similar technical measures impossible or uneconomical.

In the nuclear power plant the use of a multi-barrier concept is possible owing to the local confinement of the radioactive substances. On the other hand,
in the process plants under consideration multiple barriers are exceptional and
even then only cover parts of the installation, since their technical realization is
extremely difficult. The necessity to ensure the integrity of the barriers against
releases of radioactive substances from the nuclear power plant leads to complex safety systems in comparison to those of the process plants considered.
In addition, the safety systems of the nuclear power plant are characterized by
a high degree of automation. This, in general, implies a greater reliability of
the system functions and less reliance on human performance. Its realization
requires comprehensive and detailed analyses of all conceivable processes in the
plant, including possible incidents right from the design stage.

In addition to design differences the various properties of the dangerous materials involved need to be taken into consideration. In nuclear power plants
radiotoxicity is the main concern. It cannot be compared directly with the

chemical toxicity of the dangerous materials used in the process plants in question.

4.4.2.4 Initiating and Undesired Events

Many more initiating events are conceivable in process plants than in nuclear power stations. They strongly depend on the type of plant and are not so well documented. This makes it necessary to apply specific methods like HAZOP Studies (cf. Sect. 4.1) in order to identify initiating events and trace their consequences in the system. In contrast to nuclear power plants, the frequencies of occurrence of many undesired events in process plants can be derived directly from statistics without requiring fault tree analyses. Since, in general, the safety systems in nuclear power plants are more complex than in process plants, the fault tree analyses for determining the unavailabilities of safety systems in [11] are considerably more detailed than those in [3].

4.4.2.5 Data for Quantifying Undesired Events

In [3] comments were made by the industrial companies involved on the reliability data used, which are summarized below:

– The choice of data is not sufficiently justified.
– It is assumed, pessimistically, that all component failures will be dangerous failures.
– The failure rates for hoses, valves, analysers, and detectors from one of the references are too high.
– It is not clear how possible dependent failures are treated.
– An entire trip system must not be described by a single probability value, because such systems are of different designs; therefore, probabilities valid for one type of system may not be applied directly to another.
– The failure rate data used for vessels are inconsistent.
– Data were used without adaptation to specific situations.

Compared to the DRS-A, less effort was made to determine component failure rate data. Systematic evaluations were not carried out. The adaptation of data to specific situations using environmental factors is questionable because the operating conditions under which the failure rates were observed are not well enough known. In the DRS-A few data were estimated by experts, whereas this was done more widely in the Rijnmond study.

Consequential events, which are important in process plant complexes, were insufficiently quantified. Common cause effects were not treated in [3] as opposed to [11]. Whereas analyses of nuclear power plants can rely on a relatively well established data base which is continuously being improved by evaluating operating experience [13], sufficient information on reliability data in the process plant environment, as for example that of [14], was not available at the time of the execution of the Rijnmond study.

4.4.2.6 Fault Tree Analyses

A comparison of the quality of the fault trees in both studies is not possible. The fault trees of [3] cannot be reviewed, since the investigated plants are not described in sufficient detail. Apart from that it would be difficult to decide what degree of detail of fault tree analysis is adequate for the study objects, since these differ substantially from one another.

4.4.2.7 Release Processes

In [3] the physical models mentioned in Sect. 4.3.4.4 were used to simulate the processes from the occurrence of the undesired event (failure of containment) to the onset of damaging effects caused by the spreading of the cloud of dangerous substances. In this way appropriate values for the quantities and types of release were obtained. This step of the Rijnmond study corresponds to the determination of the release categories in the DRS-A. It has to be borne in mind, however, that the German risk study does not address any hazards to employees and that damage to the population can only occur after spreading of the cloud of dangerous substances.

In the case of late containment failure (cf. Sect. 3.1.1.4) radioactive decay and deposition processes inside the containment will reduce the quantity of radioactive substances released to the environment. The process plants investigated, however, have no containment and toxic substances do not undergo transformation similar to radioactive decay so that the entire quantity released is transported into the environment. Therefore the time factor is important with the DRS-A but not in the Rijnmond study. In both investigations model calculations of the release behaviour of the dangerous materials have considerable influence on the final results of the studies.

In the Rijnmond study distinct models were used to describe the variety of damage (fires, explosions, or toxic damage) and to account for the different physical and chemical states which influence the release behaviour of the dangerous substances (cf. Sect. 4.3.5). Differences in the quality of these model calculations impede the comparison of results of different release scenarios.

With the DRS-A all accident scenarios were treated with a single model. The differences in scenarios were accounted for by various boundary conditions for the model calculations. This is possible in the case of late containment failure, due to the formation of a "containment atmosphere", making the thermodynamic states for the different accident scenarios similar.

4.4.2.8 Dispersion Calculations

Due to the different boundary conditions for releases and the necessity to also treat gases which are denser than air, two dispersion models were used in the Rijnmond study. One model, which accounts for the different failure modes by appropriate boundary conditions, is sufficient to describe the releases from the containment of a nuclear power station. In both cases meteorological data were taken from weather statistics. Additionally, the dispersion parameters were measured especially for the DRS-A and are site-specific. Furthermore, sensitivity analyses were carried out in order to assess the influence of uncertainties of the meterological parameters. This was not done in the Rijnmond study.

4.4.2.9 Damage Models

Because of the great variety of substances used in the process industry and of the types of damage caused by them it is more difficult to determine appropriate dose-effect relationships than in the nuclear field. In addition, the effects of ionizing radiation have been investigated for a long time, while the toxic effects of many chemical substances are still widely unknown. This is also reflected in the industrial comments in [3], in which, among others, the lack of appropriate observations on the effects of toxic substances on the human organism is underlined.

4.4.2.10 Summary

The Rijnmond study was meant to be a pilot study with the chief objective of gaining experience in applying risk analysis methods to plants of the process industry. The individual and collective risks for employees and the population in the vicinity of an industrial complex were to be assessed. A well-defined part of the industrial complex was chosen as an example for this purpose.

The German risk study, on the other hand, was primarily to determine the societal risk from accidents in nuclear power stations. Based on the results obtained for a reference plant, the collective risk for the entire population was calculated, which could result from the operation of 25 such plants located at 19 different sites in the Federal Republic of Germany. Apart from the fact that the objectives of both studies are different there are considerable differences in procedures, applied methods, completeness, and credibility of results. These are summarized below:

- The process plants investigated differ from nuclear plants with regard to design, type and local distribution of damaging substances, operational and safety concepts, etc. The multibarrier concept and the compact design of nuclear power plants is in contrast to the open type of structure generally used in process plants and their extension over considerable areas. Such differences imply distinct procedures for identifying initiating events and for analysing risk-relevant event sequences. They do not necessarily impede the comparability of the results. However, it is more difficult to judge whether all risk-relevant event sequences have been considered in the study of the process plants in question.
- The quality of the reliability data bases differs and affects the comparability of the results, particularly with respect to quantifying uncertainties. In the Rijnmond study engineering judgement had to be used more frequently than in the DRS-A to obtain reliability data because of the scarcity of statistical

information. For example, the adaptation of data to specific situations was carried out using estimated environmental factors. In addition, the insufficient quantification of consequential events and the neglect of common cause influences in the Rijnmond study impedes any comparison of the results.

- The influence of the differences between the models for release processes (discharge, evaporation, etc.) in the Rijnmond study and the DRS-A on the comparability of the results cannot be assessed. Some of the models are complex and the physical processes treated are different. Comparison therefore is difficult. This also applies to the calculation of the atmospheric dispersion of dangerous substances, since in the case of radioactive releases much larger areas have to be considered than for toxic substances. Furthermore, the site-averaged assessment of the DRS-A requires a basically different approach to modelling meteorological influences.

- Differences impeding the comparison also exist in the determination of health effects. The large number of substances used in the chemical industry and the different damage-causing mechanisms make the determination of dose-effect relationships more difficult than for radioactive substances. For this reason much less is presently known on the toxic effects of chemical substances than on the health effects of radioactivity. Thus, for example, possible carcinogenic effects of toxic substances were not taken into account in the Rijnmond study and consequently late health effects were not assessed. Possible genetically significant effects were also not considered.

In summary, it can be concluded from the aforementioned differences that the results of both studies are not comparable. If risk studies for process plants are to reach the state of development of risk studies for nuclear power stations, the following requirements must be satisfied:

- Systematic collection of reliability data for components in process plants under adequate quality assurance.
- Systematic evaluation of operating experience in order to obtain reliable information on accident-initiating events and their frequencies of occurrence.
- Development and validation of models for physical and chemical processes like the discharge of multi-phase and multi-component mixtures and for vapour cloud explosions.
- Development of models for the atmospheric dispersion of explosible and toxic substances, accounting for surface roughness and buildings in the vicinity of the location of release.
- Determination of dose-effect relationships for important toxic substances.

References

1. Lees, F.P.: Loss prevention in the process industries. Vols. I and II. London 1980
2. Canvey – An investigation of potential hazards from operations in the Canvey Island/Thurrock Area. London 1978

3. Risk analysis of six potentially hazardous industrial objects in the Rijnmond Area. A Pilot Study. Dordrecht, Holland / Boston, U.S.A. / London, England 1982
4. Menzies, R.M.; Strong, R.: Some methods of loss prevention. The Chemical Engineer, p.151, March 1979
5. The Dow Chemical Company: Fire and explosion index – hazard classification guide, May 1987
6. Ausfalleffektanalyse. DIN 25448 (1980)
7. ICI Ltd.: Hazard and operability studies. Process safety Rep. 2. London 1974
8. Ereignisablaufanalyse – Verfahren, graphische Symbole und Auswertung. DIN-25419 (1985)
9. Canvey – A second report. A review of potential hazards from operations in the Canvey Island/Thurrock area three years after publication of the Canvey Report. London 1981
10. Methods for the calculation of the physical effects of the escape of dangerous materials (liquids and gases). Bureau for Industrial Safety T.N.O. Directorate General of Labour. Voerburg, The Netherlands 1979
11. Deutsche Risikostudie Kernkraftwerke: Eine Untersuchung zu dem durch Störfälle in Kernkraftwerken verursachten Risiko. Köln, 1979 (English translation: German risk study – main report. A study of the risk due to accidents in nuclear power plants. EPRI NP-1804-SR. Palo Alto, California, April 1981)
12. Reactor study – an assessment of accident risks in US commercial nuclear power plants. WASH-1400/NUREG-075/014, 1975
13. Hömke, P. et al.: Zuverlässigkeitskenngrößenermittlung im Kernkraftwerk Biblis B; Abschlußbericht. GRS-A-1030, Köln 1984
14. Doberstein, H.; Hauptmanns, U.; Hömke, P.; Verstegen, C.; Yllera, J.: Ermittlung von Zuverlässigkeitskenngrößen für Chemieanlagen. GRS-A-1500, Köln 1988

5 Risk Comparisons for Nuclear and Conventional Energy Conversion Systems

The debate on the risks of the peaceful use of atomic energy has intensified public awareness of impacts on the environment. It is increasingly understood that almost all human activities may adversely affect the environment, at least if they are carried out on a large scale. This is true even for energy conversion options which for a long time have been considered free from risk, e.g. the use of solar or wind energy. Therefore, it is reasonable to determine the impacts resulting from the use of different energy sources. They should be compared with one another and accounted for when making decisions on how to satisfy the energy demand of a country. It must be kept in mind, however, that risk is only one aspect among many others in the assessment of an energy supply system. According to [1] the following criteria should be considered:

- economy,
- international compatibility,
- environmental compatibility,
- societal compatibility.

The interrelations between the different aspects should be accounted for. This is, however, impossible given the present state of knowledge, particularly, if the degree of fulfilment of the individual criteria is to be quantified.

5.1 Problems, Benefits and Bases of Comparison

According to [1] the comparison of possible detrimental effects of the use of different energy sources can presently not be based on comprehensive scientific investigations. It is emphasized that there are problems in assessing the impacts of non-nuclear energy sources and that even bigger gaps exist in appraising renewable energy sources. On the other hand, risk comparisons may be important in the treatment of a variety of problems, as is shown in Table 5.1, which is taken from the extensive study in [2]. In particular, risk comparisons may be one of the possible bases for formulating safety goals for nuclear power plants (cf. Sect. 6.6).

A prerequisite for comparing risks is the use of clearly defined notions (cf. Sect. 1.1). In addition, the choice of the basis of comparison requires special

Table 5.1. Summary of uses, users, benefits and interchangeability of electrical power sources (from [2])

A. Uses of analyses

1. Site specific studies
   Utility planning studies
   Power grid planning studies
   National energy supply planning
2. Global planning
   International energy planning
3. Public education and information dissemination
4. Special purpose analyses
   Energy sub-systems investment
   Evaluation of potential problems in new energy sources
   To support or reject an energy option

B. Users of studies

1. Policy formulation users
   Utilities
   Government (local and regional)
   Energy suppliers and related industries
   National and international authorities
2. Technical community
   – scientists, engineers, economists
3. General public

C. Benefits of electrical energy production

1. Benefits of particular choices of energy sources
   Continuity of supply
   Preservation of national resources
   Vulnerability
2. Economics and fuel supply
   Balance of payments
   Investment strategies
   Economies of scale
   Price elasticity of demand for fuel sources
   Reliability and capacity
   Substitutability
3. Indirect benefits of energy production
   Employment
   By-products of energy production
4. Opportunities for efficient use of energy mixes
   Distributed and centralized energy source mixes
   Matching needs for base loads and variable loads
   Plant capacity
5. Negative benefits

D. Interchangeability of energy sources

1. Supply attributes of alternative sources of energy
2. End use forms of the energy
3. Site location conditions

---

attention. For example, a power station of 1000 MWe may serve as a reference; the corresponding number of units would have to be taken if the plants were smaller. A different choice may be the total energy generation of a plant during its lifetime. Another could be the services rendered to the end user; still another the total risk of an entire electrical power grid. In this case the different areas of application of the various processes for base, medium, and peak load would be accounted for. It is evident that the choice of the basis of comparison will influence the calculated risk.

In the following a survey of investigations of the risks of different energy sources is given; the methods of analysis and the results are critically examined.

5.2 Technical Parameters and Data for Environmental Impact of Energy Conversion Systems

5.2.1 Preliminary Remarks

Table 5.2 contains a list of the evaluated investigations. The energy conversion systems are listed according to increasing risk. The ranking is based on the number of fatalities. No differentiation is made between early and late fatalities.

Since the results of the risk evaluations depend on the technical parameters of the systems under investigation, these are given in Tables 5.3 to 5.6. For comparison, the corresponding values from a data collection performed for IEA [19] are listed as well. Many of the reviewed studies do not indicate the technical parameters of the systems considered, or, at best, state references. Hence, it is difficult to reproduce the results. In these cases, entries in Tables 5.3 to 5.6 are missing. Some of the studies examine systems whose technical maturity can only be expected in the future, as for example the use of heliostats or of

-----

[a] In addition electricity generation from lignite is treated.
[b] Ranking based on fatalities, as stated in [6]; it changes if back-up systems for renewable energy sources are included.
[c] Results for wind energy are calculated according to the input-output method (cf. Sect. 5.4.2); those for gas and oil were taken from the literature.
[d] The type of comparison is doubtful.
[e] Data are based on operating experience of the Ontario Hydro (Canada). Since the type of fossil energy source used is not mentioned the corresponding data were entered under coal. The authors of [11] regard the risks of the three energy conversion systems treated as approximately equal.
[f] Additionally, electricity generation from burning peat is treated.
[g] The same figures as in [7].
X: Is not used for electricity generation and therefore not included in the risk comparison.
Y: Risk is only discussed qualitatively.

Table 5.2. Evaluated studies and the energy conversion systems treated (ordered according to increasing risk; author's calculation)

Energy source / conversion process	[3]	[4][a]	[5]	[6][b]	[7][c]	[8][d]	[9]	[10]	[11][e]	[12][f]	[13]	[14][g]	[15]	[16]	[17]	[18]
Year of publication	1974	1976	1977	1978	1980	1981	1981	1981	1981	1981	1981	1979	1981	1981	1981	1981
Coal	4	4	2	10	5	2	5	Y	1	2	3	2	3			
Oil	3	3		9	4		1				1		2			
Natural gas	1	3		1	1											
Fission (LWR)	2	2	1	2	3		2		2	1	2	1	1			4
Solar (heat electricity)				6			4									
Photovoltaic energy				7			3							2		5
Solar room and water heating				X										X		
Wind				5	2	1								1		
Ocean heat				3												
Methanol				8												
Hydropower		1	Y	4					3							
Fast breeder reactor															3	
Solar energy from satellites															2	
Decentralized photoelectric energy														5	6	
Coal gasification with gas turbine															7	
Nuclear fusion															1	Y
Passive solar heating														X		
Active solar water heating														X		
Active solar heating and cooling														X		
Solar process heat														X		
Anaerobic fermentation														X		
Wood pyrolisis														X		
Decentralized wind energy with storage														3		
Decentralized wind energy without storage														4		

satellites for converting solar energy. Naturally, in these cases less confidence should be placed on the data than in the case of well-established systems like coal-fired or LWR power stations.

In considering the potential use of risk comparisons for decision-making it must be borne in mind that not all of the available systems may be used in every country. For example, solar heat for electricity generation is not economic in the FRG, because of climatic reasons [20].

The environmental impact of the different processes is briefly reviewed in the following sections; a summary is given in Table 5.7.

5.2.2   Coal-fired Power Plants

The technical parameters of the systems for the generation of electricity from coal are given in Table 5.3.

Adverse impact on the environment in this case derives from waste heat and the emission of noxious substances. According to [4] the discharge of waste heat to rivers has the following negative impact:

– decrease of the solubility of oxygen in the water of the river,
– increased consumption of oxygen due to an accelerated degradation of organic matter in the water of the river,
– alterations of the biorhythms of plants and animals,
– changes of the overall toxic situation.

These effects may be counteracted by the use of wet or dry cooling towers.

The air pollution caused by coal-fired power stations is treated in detail in [21] and [22]. The following brief discussion is mainly based on [22].

The emissions chiefly consist of the oxides of carbon, nitrogen (NO_x), and sulphur, hydrocarbons, particulate matter, trace minerals, and hydrogen halides. SO_2, hydrogen halides, and nitric oxides irritate the mucous membranes in different parts of the respiratory tract. CO blocks the haemoglobin. Polycyclic aromatic hydrocarbons (PAH) produced in the combustion process and bound to dust particles may cause cancer after inhalation. Heavy metals like arsenic, beryllium, cadmium, chromium, nickel, and antimony are also carcinogenic. They are emitted in small quantities, but possibly selectively concentrated in the environment. Some types of coal also contain radioactive substances; according to [23], however, these contribute little to the frequency of cancer. An investigation carried out in the FRG [24] supports this view; it estimates an additional radiation exposure resulting from the operation of all coal and nuclear power plants in the FRG of 1% of the natural radiation exposure. The emission of CO_2 may cause climatic problems in the long run, as explained in [22] and [25].

Summing up, it is stated in [22] that the necessary basis for a risk comparison between nuclear fission and the combustion of coal is still lacking, particularly as

Table 5.3. Technical parameters and data on environmental impacts for the generation of electricity on the basis of coal

Study Parameter	Energy source: coal				
	[19][a]	[3]	[4]	[12]	[6]
Power in MWe	2 · 649	1000		2 · 500	1000
Annual energy input in 10^{15} J	43.4	62.2			
Efficiency in %	34	38			37
Annual energy production in 10^{15} J	14.8	23.7			
Availability in %	85				
Load factor in %	45.7	75	70		70
Technical life in yr	30				30
Surface requirements in km^2/GWe	0.5	1.6			
Particulates in g/KWh	1.26	0.3[c]	0.49	0.19	n.i.
NO_x in g/KWh	2.59	4.1			n.i.
SO_x in g/KWh	9.68	3.65[b,c]	7.98[c]	5.4[c][d]	n.i.
CO in g/KWh	0.17	0.15			n.i.
$C_n H_m$ in g/KWh	0.04				n.i.
Radioactivity in Bq/KWh	unknown	0.11			n.i.

[a] Mean values without the error bounds indicated in [19]; year of reference: 1978.
[b] Only SO_2
[c] With abatement.
[d] Additionally $0.035 - 0.35 \cdot 10^{-3}$ benzopyrene.
n.i.: not indicated.

Table 5.4. Technical parameters and data on environmental impacts for the generation of electricity on the basis of oil and natural gas

Study Parameter	Energy source				
	oil			natural gas	
	[19][a]	[3]	[4]	[19]	[3]
Power in MWe	2 · 600	1000	1200	2 · 600	1000
Annual energy input in 10^{15} J	76.5	60.6		37.9	62.2
Efficiency in %	37.2	39.0		34	38
Annual energy production in 10^{15} J	28.6	23.7		12.9	23.7
Availability in %	85			85	
Load factor in %	75	75	70	34	75
Technical life in yr	30			30	
Surface requirements in km^2/GWe	2.3	0.32		2.2	0.4
Particulates in g/KWh	0.33	0.023[b]	0.16	≈0	0.006[b]
NO_x in g/KWh	2.3	1.3		1.8	2.0
SO_2 in g/KWh	7.6	3.2[b]	7.96	≈0	0.0027[b]
CO in g/KWh	0.03	1.14		0.04	0.0018
$C_n H_m$ in g/KWh	0.08	0.14		–	0.17
Radioactivity in Bq/KWh	–	0.003		–	Negligible

[a] Mean values without the error bounds indicated in [19]; year of reference: 1978.
[b] With abatement.

Table 5.5. Technical parameters and data on environmental impacts for the generation of electricity on the basis of solar energy and nuclear fission in pressurized water reactors

Parameter	Study Energy source Fission (PWR) [19][a]	[3][b]	Solar heat [19][c]	[6]	Photoel. conv. [19][d]	[6]	[17]
Power in MWe	1228	1000	100 peak	1000	1 peak	1000	200
Annual energy input in 10^{15} J		74.1	3.528		0.0552		
Efficiency in %	32.6	32.0	15.3		12		
Annual energy production in 10^{15} J	29.0	23.7	0.54		0.0066		
Availability in %	75		85		85		
Load factor in %	75	75	17	70	21		25.8
Technical life in yr	30		25	30	25	30	30
Surface requirements in km^2/GWe		0.4	10.5		10		
Radioactive emissions in 10^{10} Bq/MWa							
Kr-85	2.6	27.0	n.a.	n.a.	n.a.	n.a.	n.a.
To the atmosphere	9.3	3.7	n.a.	n.a.	n.a.	n.a.	n.a.
To water	3.3	2.8	n.a.	n.a.	n.a.	n.a.	n.a.

[a] Mean values without the error bounds indicated in [19]; year of reference: 1978.
[b] Some indications omitted.
[c] Mean values without the bounds indicated in [19]; year of reference: 1985.
[d] Mean values without the bounds indicated in [19]; year of reference: 1990.
n.a.: not applicable.

Table 5.6. Technical parameters and data on environmental impacts for the generation of electricity on the basis of wind energy, water, and nuclear fission in fast breeder reactors

Parameter	Study Energy source Wind [19][a]	[6]	Water (reservoir) [19][b]	[6]	Fast breeder [19][c]	[14, 17]
Power in MWe	3	1000	240	1000	1200	1250
Annual energy input in 10^{15} J	0.307		3.0			
Efficiency in %	14		76		40	
Annual energy production in 10^{15} J	0.043		2.3		28.4	
Availability in %	95		95		75	
Load factor in %	46	20	30	60	75	70
Technical life in yr	25	20	60	50	30	
Surface requirements in km^2/GWe	0.33		6.5			
Radioactive emissions in 10^{10} Bq/MWa						
Kr-85	n.a.	n.a.	n.a.	n.a.	3.7	n.i.
To the atmosphere	n.a.	n.a.	n.a.	n.a.	26	n.i.
To water	n.a.	n.a.	n.a.	n.a.	0.37	n.i.

[a] Mean values without the error bounds indicated in [19]; year of reference: 1985.
[b] Mean values without the error bounds indicated in [19]; year of reference: 1978.
[c] Mean values without the error bounds indicated in [19]; year of reference: 1995.
n.i.: not indicated.
n.a.: not applicable.

Table 5.7. Important environmental impacts from the use of different energy sources (from [2])

	Hydropower	Coal	Oil	Natural Gas	Liquefied Gas	Nuclear	Renewable energy sources	Synthetic fuel	Other fossil supplies
Mining and extraction									
Workers	–	Large accidents, Pneumocosis	Medium (offshore) accidents	–	–	Small accidents, Radon	Maintenance accidents	–	–
Public	–	Accidents, Mine drainage	Oil spills	–	–	Tailings piles, Chemicals	–	Water demand	Large area of disruption, Water demand (for oil shale)
Preparation	Changes in living patterns	–	Refining wastes and effluents	–	–	–	–	Carcinogenic chemicals, Waste	Waste volume
Transportation of fuels									
Workers	–	Low	High	Medium	High	Low	–	High	High (depending on fuel type)
Power generation	Dam failure leading to floods	Thermal loading	Thermal loading	Thermal loading	Thermal loading	Thermal loading, Accidents of various sizes	Extensive land use, Interruptible	–	–
Waste management									
Effluents	–	Particulates SO_x, NO_x, CO_x	SO_x, NO_x, CO_x	CO_x	CO_x	Radioactive effluent	–	NO_x, CO_x	NO_x, CO_x
Wastes	–	Fly ash Culm piles	–	–	–	Radioactive waste	–	–	–

far as coal is concerned. In [7] the absence of dose-effect relationships, especially for low concentrations of air pollution, is pointed out. For this reason a linear dose-effect relationship without threshold as in the analysis of nuclear power stations (cf. Sect. 3.1.2.3) is used.

In addition, there are indications, though without scientific proof, that sulphuric and nitric oxides and their transformation products significantly contribute to damage to forests, waters, and buildings [26]. This problem is now receiving international attention under the heading "acid rain" [27]. Property damage associated with this phenomenon may be estimated, but the loss of quality of life is not quantifiable and may, therefore, not be included in any calculation of risks.

5.2.3 Oil-fired Power Plants

The technical parameters for generating electricity from oil are shown in Table 5.4. The statements made in the previous section about the waste heat problem apply here too.

The emissions from power stations which burn oil consist of oxides of carbon, nitrogen and sulphur, and dust particles; the amount of nitric oxides is similiar to that from coal-fired power plants. The emission of particles reaches only 1/100 of the quantity indicated for coal. In addition, there are small quantities of hydrocarbons and, depending on the origin of the oil, also traces of metals [3]. As far as damage to health is concerned, in essence the remarks made in the previous section apply.

5.2.4 Natural Gas-powered Plants

The technical parameters for generating electricity on the basis of natural gas are given in Table 5.4. The waste heat problem is similar to that of coal-fired power stations. The same is true for the emissions of noxious substances, which also consist of dust particles, and oxides of carbon, sulphur and nitrogen. The quantities, however, are much smaller. Natural gas is probably one of the cleanest fuels [2]. As far as damage to health is concerned, the remarks of Sect. 5.2.2 apply, mutatis mutandis.

5.2.5 Nuclear Power Plants

The technical parameters for generating electricity with pressurized water reactors are shown in Table 5.5. With this technology the waste heat problem addressed in Sect. 5.2.2 also exists, but owing to the lower efficiency the quantity of waste heat per unit of electric energy produced is larger than for fossil plants. Environmental impact from nuclear power stations derives from the biological effects of the radioactive isotopes built up during operation. In addition to the normal operation, and in contrast to the previously treated systems, the possibility of major accidents has to be taken into account. These may be accompanied by large releases of radioactive substances, whose health effects are explained in Sect. 3.1.2.3.

Cases of acute radiation sickness cannot occur if legally fixed dose limits are not exceeded. Stochastic health damage, however, may not be excluded, but its proportion is below that resulting from the exposure to natural radiation.

In [28] the complementary cumulative frequency distributions shown in Figs. 3.10 and 3.11 for early and late fatalities as a consequence of the operation of 25 plants in the FRG were calculated. The assumed total power was 32 500 MWe, whereas the present total installed capacity of all nuclear plants in the FRG is 23 952 MWe.

Technical data for the fast breeder reactor, which is included in the risk comparison in [17], are found in Table 5.6. This type of reactor produces about the same quantity of waste heat as a conventional power station of the same capacity. Just as in case of the pressurized water reactor, impact resulting from normal operation and from accidents has to be considered. The study discussed in Sect. 3.3.1 showed that accidents in the fast breeder reactor analysed there involve no greater risk than those in a pressurized water reactor.

A further system treated in [17] is the fusion reactor. Since controlled fusion so far has not even been achieved at laboratory level dependable statements about environmental impact are not yet possible due to lack of a definite technical plant concept. According to [29] the following aspects have to be taken into account when assessing the risk.

During normal operation the radioactive inventory of a fusion reactor is largely contained inside the reactor. In contrast to fission reactors, the reprocessing of spent fuel in special centres and the final storage of the radioactive waste are not necessary. This is, however, not true for the structural materials of the reactor which are activated during operation and have to be stored safely. In case of an accidental release of tritium, the radioactive exposure of the environment is expected to be two or three orders of magnitude below that from an accident in a fast breeder reactor of equal power.

5.2.6 Renewable Energy Sources

Technical data for some of the energy conversion systems on the basis of renewable energy sources are shown in Tables 5.5 and 5.6. The environmental impact of

- solar energy,
- biomass (combustion and gasification of wood and straw),
- biogas,

- wind power,
- hydropower.

is described in [21, 22]. The following remarks are based on these references.

The use of renewable energy sources is generally considered to have no adverse impact on the environment. However, this is not true, because the energy density of these sources is small compared with fossil fuel or nuclear energy. They often require large structures for a comparable quantity of energy to be produced. This means much material and land has to be used. Additionally, the following aspects should be taken into account.

In general, negligible air, water or solid pollutants result from the use of solar energy. However, accidental or emergency releases and periodic system flushing from solar thermal units may release pollutants depending on the fluids involved. Heliostat fields might affect local ecology, especially if the plant is located in desert areas. In the case of solar photovoltaic power plants there is a potential for accidental high voltage arcing, which will generate ozone and fumes from burnt plastics. On decommissioning solid wastes such as concrete, silicon and the glazing must be disposed of.

Compared to fossil fuel combustion, in burning some types of biomass low levels of sulphur oxide emissions, but high levels of particulates and organic gases are produced. However, a complete combustion, especially in medium and large installations, reduces the emissions considerably. The main emission problems occur with incomplete combustion in small installations.

Energy generation with windmills is accompanied by "visual pollution", noise, television interference, and breakdown hazards (e.g. the blade may come off its supporting structure or the entire windmill may topple over).

The use of run-of-the-river, pumped storage and reservoir hydropower plants may cause changes in the quality of water, of the ecology of the rivers, and of fish migration. The construction of dams radically converts terrestric ecosystems into aquatic ecosystems. Yet, in relation to an entire country the ecological damage is small. There exists, however, the possibility of catastrophic dam failures. This is treated in [21, 30, 31].

The cultivation of fast growing trees (e.g. certain kinds of poplars) for the production of methanol would reduce the living space for wild plants and animals. The necessary fertilization severely interferes with the natural ecosystem. Considerable loads on the aquatic system must be expected from the washing out of nutrients.

Finally, some remarks are made on the environmental impact of geothermal energy conversion, although this system is not treated in any of the reviewed risk comparisons. According to [32] the environmental impact from this energy conversion is less than that caused by other types of thermal power stations. Nevertheless, especially in the case of large plants, considerable emissions of hydrogen sulphide and CO_2 are to be expected. The specific quantities, however, are in general lower than those for coal-fired power stations. In addition, there may be noise emission.

5.3 Environmental Impact from the Fuel Cycle

5.3.1 Preliminary Remarks

The operation of power stations using non-renewable energy sources requires a fuel cycle which also affects the environment. This has to be taken into account, especially if a comparison with renewable energy sources is intended. It must be kept in mind, however, that the impact is not necessarily produced within the boundaries of the economic area under consideration. For example, the mining, extraction, and preparation of the fuel may take place in other countries. This should, however, not be a justification for neglecting their effects.

In what follows the environmental impact of fuel extraction, preparation, and processing is treated. Transportation risks are mentioned in some cases, occupational risks are not addressed. However, both are outlined in Table 5.7 and contained in the figures of Tables A1-A15 of the Appendix; a detailed account may be found, for instance, in [21].

5.3.2 Coal

According to [22], releases of methane and acid mine drainage are to be expected in coal mining. In addition, noise emissions and subsidence caused by mining may be considered as an environmental impact from coal extraction. During the preparation of coal, waste is separated and dust and waste water, which requires cleaning, are emitted. In addition, the mine spoil, which is frequently tipped, and the coal and coke piles occupy considerable space.

The extraction of lignite, which mostly takes place as opencast mining, interferes with the natural water flood process. This affects wells in the surroundings. Subsidence may occur, and the vicinity of the mines may be affected by noise and dust emissions. The substantial land use may be compensated in part by appropriate reclamation after exploitation.

5.3.3 Natural Oil and Natural Gas

Adverse environmental impact from the exploitation of oil and natural gas comes from chemical substances which are recovered in the extraction process, e.g. hydrogen sulphide and brine with oil and hydrogen sulphide and carbon dioxide with natural gas. In addition, there is always the possibility of explosion when oil or gas are found. The offshore oil exploitation may cause water contamination and be accompanied by catastrophic accidents at oil drilling platforms. Transportation may be an ecological threat because of possible spills.

The operation of a refinery which converts the crude oil to gasoline, fuel, and lubricants causes emissions of organic gases and vapours, hydrogen sulphides, sulphur dioxide, mercaptanes, and dust. In addition, noise is emitted. Effluents are no longer a problem with the present state of technology, but separated noxious solids have to be stored.

In processing natural gas H_2O, and with certain types of gas, SO_2 are emitted to the environment as flue gases. The land required for a refinery is considerable.

5.3.4   Nuclear Energy

The fuel cycle for nuclear power plants is treated in detail in Sect. 3.4. In the case of the light water reactors usually employed it comprises the following steps:

– uranium mining, extraction, and preparation
– uranium enrichment,
– fuel fabrication,
– reprocessing and fabrication of mixed oxide fuel elements,
– final storage.

According to [22] uranium mining, extraction, and preparation do not have a global impact on the environment. This is also true for the normal operation of enrichment plants. However, there is the possibility of a release of UF_6 whose toxicity is more of a hazard than its radiation effects.

The emissions from plants for converting enriched uranium into UO_2 and for the fabrication of nuclear fuels have in the past remained below the legally fixed limits. Through appropriate design it is ensured that critical masses, which might lead to an accident, cannot be formed. The environmental impact from this part of the fuel cycle is small.

The fabrication of nuclear fuel and the operation of nuclear power stations continuously creates radioactive wastes, primarily in form of loaded filters from air conditioning equipment, ion exchanger resins, and residuals from evaporation processes. In addition, spent fuel elements and radioactively contaminated or activated parts of decommissioned power stations have to be disposed of.

Conditioning and transport of waste – with the exception of spent fuel elements – does not cause significant impact on the environment. There is still insufficient experience with dismantling nuclear power stations.

Spent fuel elements in the FRG are at present stored in interim storages. These are located within the power plant itself or at specific sites, for example, Ahaus (under construction) and Gorleben (ready for operation). In [28] the risk of storing spent fuel elements inside the power station is treated. It is considered small in comparison with the total risk of the plant.

The risk of the release of radioactive substances during the transport of spent fuel elements to reprocessing plants or direct final storage, i.e. without reprocessing, is regarded as small because of the safety measures taken [22] (cf. Fig. 3.27 and Table 3.21).

The emissions of the reprocessing plant at Karlsruhe (FRG), which has been in operation since September 1971, have always remained well below the admissible values. However, in [22] the view is taken that for plants of larger capacity the retention efficiency should be improved. This is technically feasible. The accidents which may occur in reprocessing plants are expected to cause only small additional releases to the environment (cf. [33] and Sect. 3.4.3.2).

The materials not used again after reprocessing and the transuranium elements must be kept away from the biosphere for a very long time. This is also true, if the spent fuel elements are stored directly. According to the present concept in the FRG it is planned to store waste with low heat production in the former iron ore mine Konrad. Waste generating large quantities of residual heat is to be stored in the salt mine at Gorleben. In this way the hazard to the biosphere is considered to be negligible. According to [34] this is true even in the case of the worst credible accident in a salt dome storage, the ingress of water.

5.4 Risk Comparisons

5.4.1 Preliminary Remark

The environmental impact reviewed in the previous section primarily, although not exclusively, constitutes a risk to the population not directly concerned with the energy conversion process. In addition, those directly involved in energy conversion are exposed to specific risks. Among them are accidents which may occur during the mining of raw materials, transport, in the power station itself, or during reprocessing and waste disposal. Occupational diseases may result from adverse conditions at the place of work, for instance, silicosis among coal miners. In addition, the construction of plants may cause accidents. In this context it must be borne in mind that in general the volume of construction increases as the energy density of the exploited energy source decreases. For this reason, the construction of the plants together with the required materials represents an important risk contribution for renewable energy sources like solar or wind energy.

5.4.2 Calculational Procedures

The total risk of a process for energy production comprises the risk of the conversion process, a possibly required fuel cycle, and the risk derived from the construction of the plants and from the acquisition of the necessary materials. Other aspects, for example, the risk of an insufficient energy supply as a consequence of renouncing the building of power stations, is not addressed in any of the studies reviewed.

As explained in Chap. 1, the assessment of risks of mass phenomena, for example, from occupational or transport accidents, is based on statistical evaluations of records on the past; for rare events, like core melt in a nuclear reactor, analytical methods are used. Damage to health as a consequence of emissions is estimated by epidemiological investigations and extrapolations from higher to lower doses. In calculating the risk contributions from the installation of plants two different methods are used: process analysis and input-output analysis.

In a process analysis the most important materials for building the plant are identified and their necessary quantities are calculated. Drawing upon statistics for the industrial branches concerned, the risk caused by their production is assessed. This is augmented by the risks from accidents during the transport of the building materials and the construction. The procedure, which is schematically shown in Fig. 5.1, is the basis of the results of [4, 6, 11].

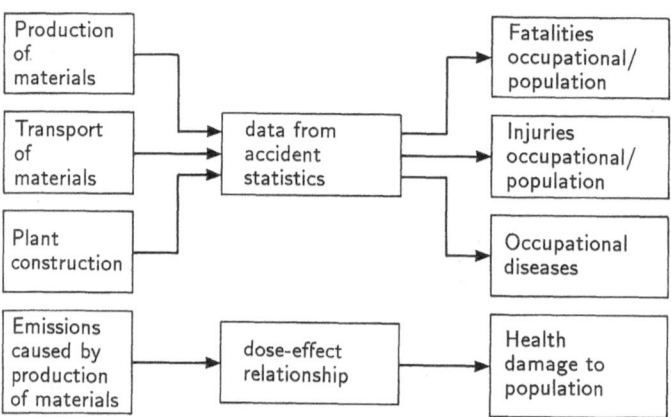

Fig. 5.1. Outline of the calculation of plant construction risks using a process analysis

In an input-output analysis [35] the interactions of all segments of an economy are represented in the form of a matrix, as shown in the example of Table 5.8. The lines contain the sales of the respective segments to all others; the columns state the input of goods and services to one segment of the economy from all the other segments. In this way, all goods and services received from any of the segments of the economy are registered in monetary units. Using appropriate factors these may be converted into risk numbers; for example, accidents per unit quantity of product for the different segments of the economy.

The application of the input-output analysis to the assessment of risks is outlined in Fig. 5.2, which shows the model developed at the Brookhaven National Laboratory (BNL) for the calculation of the environmental impact and health effects of different technologies. In this model the economy is divided into 110

Table 5.8. Example of an input-output table (as cited in [35])

Output Input	Agriculture and fishing	Ferrous metals	Non-metallic minerals	Construction
Agriculture and fishing	0.076	0	0	0.008
Ferrous metals	0.006	0.305	0	0.152
Nonmetallic minerals	0.005	0.010	0.101	0.530
Construction	0.026	0.004	0.002	0

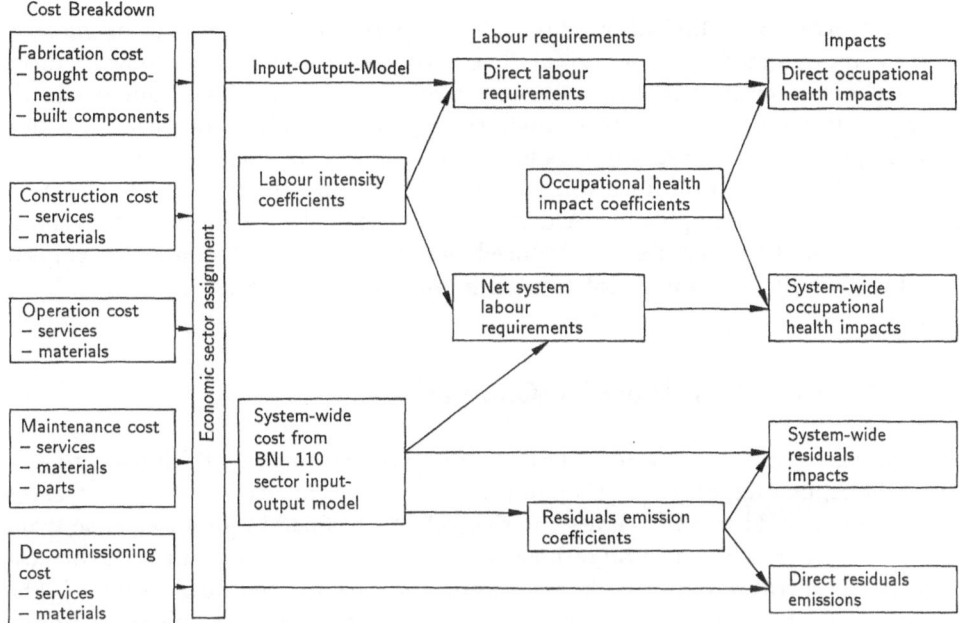

Fig. 5.2. Outline of the Brookhaven National Laboratory (BNL) model for calculating environmental impacts and health effects (from [16])

segments. The advantage of using the input-output analysis stems from the fact that the contribution of the entire economy to the construction of a plant is accounted for. Therefore there is no need to determine which materials and activities contribute substantially to risk, as is required for process analysis.

As a practical example for the difference between process analysis and input-output analysis figures are given which were obtained in [7] for deaths and injuries in the construction of windmills. If the contributions of the total economy (system-wide effects) are taken into account, the risk is estimated by a factor of two or three higher than if the analysis is limited to the directly used materials and the construction process of the plant. Further details are found in Table 5.9.

Table 5.9. Labour requirements and occupational accidents for generating $1.06 \cdot 10^{15}$ J using windmills calculated according to the process and input-output analyses (preliminary values [7])

	Fatalities in 10^{-4}	Illness in 10^{-1}	Labour in man-years
Direct occupational impacts (process analysis)	78.5	55.1	118
System-wide occupational impacts (input-output analysis)	171.0	196.0	243

Nevertheless, a limitation of both procedures is that all changes in the economy are assumed to be marginal. The activities of building the power station must not interfere with the economy so as to cause significant changes in the production and accident structure in the segments involved. However, this does not apply to major programmes for promoting certain energy systems. In such cases new production lines will have to be built and the figures from the past will no longer be applicable. A detailed analysis of the economy would then be required and the risk data to be used would have to be estimated by experts. This difficult situation is not treated in any of the studies.

5.4.3   Results of the Risk Comparisons

The risks estimated in the reviewed studies are presented in Tables A-1 to A-14 of the Appendix. Since in [16] and [17] some decentralized systems are treated, Table A-15 additionally shows the risk values obtained in [6] for electric transmission lines. The question of how centralized systems are compared correctly with decentralized ones is, however, not addressed. According to [6] a higher risk might be expected for decentralized plants compared to centralized ones.

All studies use expected values for representing risks. The effects of catastrophic events (e.g. possible releases of large quantities of radioactive substances) which should be represented in the form of complementary cumulative distribution functions (CCDF) (cf. Sect. 1.1.6) are not stated separately, but included in the expected values. In some investigations, occupational and population risks are assessed. Others limit themselves to the former. Occasionally details on accidental deaths and injuries and on late fatalities and invalidity are given. Other studies only indicate the consequences of accidents. If the loss of human life is converted into a loss of working time, as is done in some of the investigations, 6000 working days lost is the conversion factor. In this way, death and permanent or temporary inability to work are converted into a sole risk unit. Table 5.2 shows that only in [6] and [9] are conventional energy sources compared with renewable ones, whilst in [16] only renewable energy sources are treated. In [17] nuclear and renewable energy sources are considered but fossil power stations are omitted.

In summary it may be said that none of the studies offers a comprehensive comparison. In addition, the criteria for the choice of the systems considered are not provided. Occasionally, the supply of space heating or cooling is covered along with electric energy generation. Since the quality of both energy forms is different the question of the basis of comparison becomes important. This problem is discussed qualitatively in [16], but no satisfactory quantitative treatment is given.

Almost all investigations consider coal as the riskiest option, whilst nuclear energy (PWR) is mostly regarded as having a low risk; only natural gas ranks lower. The inclusion of renewable energy sources affects the risk ranking of nuclear energy in different ways. In [4] it remains in second place, in [7] it drops to the third rank, and in [13] it moves to the middle range. This demonstrates the importance of the choice of the competing options for a risk comparison.

The differences between numbers for fatalities calculated in the studies are shown in Table 5.10. The ranges between the maximum and minimum values are considerable. A detailed analysis of the reasons for this is not possible in the present context, nor can the results be traced to their sources, which probably are the same for many of the investigations. However, some aspects should be mentioned which can possibly explain the differences.

The technical designs of the plants that form the basis of the studies are different. This is evident from Tables 5.3 to 5.6, where the substantial difference of emissions should be noted. As can be seen from Tables A-1 to A-14, different quantities of labour are assigned to the individual process steps in the various investigations. For population risk assessment the uncertainties of the dose-effect relationship (cf. Sect. 5.2.2) are important. In some cases different processes to obtain raw materials are used in different countries (e.g. in the U.S.A. a major portion of hard coal is produced by opencast mining, which is not the case in the FRG). The state of process and safety technology is not identical in all countries and at all times. This is reflected by the accident figures shown in Table 5.11. A further reason is the different methodology (cf. Sect. 5.4.2) of risk assessment. Last but not least, the uncertainty of productivity and accident coefficients and the necessity to adapt information from other areas due to lack of data (e.g. the application of accident statistics from roofing to the installation of solar cells) play a role. Also problems of delineation are important (What is a labour accident, a transport accident, an occupational disease, etc.?). In addition, the handling of the large quantities of data involved may lead to mistakes of transcription and other inaccuracies; for example, the latter gave rise to corrections in the original publication of [6].

Table 5.10. Maximum and minimum values and ranges for fatalities from the use of different options for energy generation, as found in the reviewed literature (reference energy: $2.21 \cdot 10^{16}$ J)

Energy source/ conversion process	Employees					Population				
	Maximum value	Reference	Minimum value	Reference	Range	Maximum value	Reference	Minimum value	Reference	Range
Coal	9.8 [a]	[7]	0.02 [a]	[11]	9.78	101.3 [a]	[5]	0.51 [a]	[3]	100.8
Oil	1.4 [a]	[6]	0.026 [a]	[13]	1.374	24.5 [a]	[6]	4.2 [a]	[13]	20.3
Natural gas	0.44 [a]	[4]	0.082 [a]	[3]	0.358	59.6	[6]	19.6	[6]	40.0
Fission (LWR)	1.04 [a]	[6]	0.03 [a]	[11]	1.01	0.6 [a]	[12]	0.008 [a]	[3]	0.592
Solar (heat-electricity)	2.8	[6]	2.6	[6]	0.2	0.54	[6]	0.2	[6]	0.34
Photovoltaic energy	1.91 [a]	[6]	0.3 [a]	[17]	1.61	3.42	[6]	1.16	[6]	2.26
Wind	2.81 [a]	[6]	0.38 [a]	[16]	2.43	0.42	[6]	0.15	[6]	0.27
Ocean heat	1.33	[6]	0.98	[6]	0.35	0.23	[6]	$7.8 \cdot 10^{-2}$	[6]	1.52
Methanol	13.4	[6]	12	[6]	1.4	1.54	[6]	0.32	[6]	1.22
Hydropower	1.82 [a]	[6]	0.017 [a]	[11]	1.803	1.125	[6]	0.771	[6]	0.35
Fast breeder reactor	0.11	[17]	0.02	[17]	0.09	0.59	[17]	0.13	[17]	0.46
Solar energy from satellites	0.42	[17]	0.16	[17]	0.26	—		—		—
Decentralized solar energy	3.5 [a]	[16]	1.34 [a]	[17]	2.16	—		—		—
Coal gasification with gas turbine	1.96	[17]	0.82	[17]	1.14	48	[17]	3.4	[17]	44.6
Nuclear fusion	0.28	[17]	0.14	[17]	0.14	$6.3 \cdot 10^{-5}$	[17]	$6.3 \cdot 10^{-5}$	[17]	0

[a] Values from more than one reference were available.

Table 5.11. Deaths in deep mining per employee and year as an example for the impact of different technical conditions on accident numbers

Country/State	Year				Reference
	1950	1960	1970	1980	
FRG	$1.29 \cdot 10^{-3}$	$8.48 \cdot 10^{-4}$	$5.86 \cdot 10^{-4}$	$4.69 \cdot 10^{-4}$	[36]
Great Britain	$7.35 \cdot 10^{-4\,a}$		$2.99 \cdot 10^{-4}$		[31]
Ontario			$6.16 \cdot 10^{-4\,b}$		[37]
U.S.A.			$8.39 \cdot 10^{-4\,c}$		[37]
Inhaber study				$1.2 \cdot 10^{-3}$	[6]

[a] Mean value from 1943 to 1952.
[b] Mean value from 1971 to 1977 converted on the basis of 1540 working hours per man and year.
[c] Mean value from 1971 to 1978.

5.5 Summary and Recommendations for Further Work

The evaluation of the studies has shown that at present there exists no appropriate methodology for risk comparisons. Witness to this is the unusually intense debate [38, 39] which began immediately after the publication of the most comprehensive study [6] and continued for some time afterwards [40 – 43]. Accordingly risk comparisons do not provide reliable results at present. This view is also held in [2]. However, a number of recommendations may be formulated which could lead to an improvement of the results.

In all cases the entire electric power grid of a country should be considered. The purpose of the grid is to satisfy a certain demand, taking into account boundary conditions like load curves and economic aspects. In order to solve this task, a series of energy conversion systems are available. Their contributions depend amongst others on their technical and economic characteristics and on the climatic and geophysical situation of the country in question. To a certain extent energy options may be substituted by one another, i.e. the contribution of one energy source to the entire energy production can be varied within certain limits. In this way, different compositions of the supply structure become possible. The total risk is then determined and the least risky composition selected, a process similar to an economic optimization problem [44]. This procedure ensures that a risk comparison specific to that country is carried out and that plants already existing in the power grid are accounted for. In addition, meaningless assumptions are avoided like the supply of base load by solar energy, which in [6] led to the debatable introduction of backup systems [45]. Since many of the renewable energy sources may only be exploited efficiently by decentralized plants their possible contributions to energy supply and to the risk would then have to be evaluated separately.

All systems – including non-renewable energy sources – should be treated by the input-output analysis, since it is more accurate to assess system-wide effects, and because the results differ substantially from those obtained with the more limited process analysis. This is also shown by the investigations of the effects which the building of coal-fired [46] and of nuclear power stations [47] has on the production and employment in the FRG. Whether it is necessary to use such a fine division of the economy into segments as is done in the aforementioned model of the Brookhaven National Laboratory will have to be investigated, particularly in view of the many other uncertainties in the analysis. The standard model for the economy of the FRG, for instance, comprises 60 segments [48]. For quantifying the construction risk the entire occupational risk of this branch should not be used, as is done in the reviewed studies. This causes distortions, since the workers in question would either work in the same segment of the economy or in another place, even if the power stations were not built. Therefore it would be reasonable to use the difference between a specific and a "mean" occupational risk. More sophisticated investigations, for example, accounting for possible unemployment in case the power stations were not built, would most likely not be justified in view of other uncertainties.

The conversion of fatalities into working days lost which is used in many studies – mostly after some apologizing remarks about the ethics of the procedure – should be abandoned in favour of treating accidents without fatalities as "partial fatalities". This would avoid a discussion on the value of life [49] and also any reproach of belittling which may justly be levelled against the treatment of such a tragic event as death in terms of work hours lost.

In addition, further knowledge of dose-effect relationships, detailed labour accident statistics, and risk studies for those areas in which rare events may occur are desirable. All value judgements made in the estimation process and the uncertainties of the results should be made explicit in the studies. Since this may not be fulfilled in the short term, it would be reasonable to create – possibly at an international level – a framework of data and methods for risk comparisons in order to make the results of individual investigations comparable to one other, despite the remaining drawbacks. An attempt in this direction was made recently [50].

Considering the fact that all studies are lacking in some way, and lead to substantially different results, it would be premature to use them as a basis of political decisions. This view is also taken in [2] whilst in [51] the belief is expressed that it is impossible to make the impacts of different technologies comparable to one another. Even there, however, quantitative risk analyses are considered to be a useful support for political decision-making.

References

1. Bericht der Enquête-Kommission "Zukünftige Kernenergie-Politik" über den Stand der Arbeit und die Ergebnisse gemäß Beschluß des Deutschen Bundestages – Drucksache 8/2628. Deutscher Bundestag 8. Wahlperiode – Drucksache 8/4341. Bonn 1981
2. Rowe, W.D.; Oterson, P.: Assessment of comparative and non-comparative factors in alternate energy systems. EUR 8844 EN 1983
3. Comparative risk-cost-benefit study of alternative sources of electrical energy. WASH-1224(1974)
4. Oberbacher, B.; Hartwig, S.; Hintz, R.: Untersuchungen über die technischen, organisatorischen und gesellschaftlichen Voraussetzungen für Risikostrategien im Bereich technologischer Entwicklung – Vergleich der Gesundheitsgefährdung bei verschiedenen Technologien der Stromerzeugung. Bericht RS I2-510 321/40 – SR 30. Battelle Institut Frankfurt 1976
5. Keeny, S.M. (Ed.): Nuclear power – issues and choices. Cambridge, Massachusetts 1977
6. Inhaber, H.: Risk of energy production. AECB-1119 (1978) 4th edition, draft version without date
7. Hamilton, L.D.: Comparative risks from different energy systems: Evolution of the methods of studies. IAEA Bulletin – Vol. 22, No. 5/6 (1980)
8. Sørensen, B.: Comparative risk assessment of total energy systems. IMFUFA tekst nr. 43. Roskilde (Denmark) 1981
9. Hubert, Ph.: Occupational risk induced by construction of energy-production chains: Methodology and evaluation in the French Case. IAEA-SM-254/50 (1981)
10. Seeliger, J.; Zimmermeyer, G.: A coal industry's view on risk comparison of energy systems. IAEA-SM-254/60 (1981)
11. Hamilton, T.R.; Wilson, R.: Comparative risks of hydraulic, thermal and nuclear work in a large electrical utility. IAEA-SM-254/03 (1981)
12. Lautkaski, R.; Pohjola, V.; Savolainen, I.; Vuori, S.: A comparative assessment of the health impacts of coal, peat and nuclear power plants. IAEA-SM-254/45 (1981)
13. Fagnani, F.; Maccia, C.; Hubert, Ph.: Comparaison des impacts sanitaires des différentes filières électrogènes – Le Cas de la France. IAEA-SM-254/51 (1981)
14. Hamilton, L.D.: Health effects of electricity generation. In: International conference on health effects of energy production. AECL-6958 (1979)
15. Hill, J.: Comparison of the benefits and risks associated with the utilisation of various energy sources. J. Inst. Nuc. Eng. 22 (1981) 107-112
16. Rowe, M.D.: Assessing systemwide occupational health and safety risks of energy technologies. IAEA-SM-254/64 (1981)
17. Habegger, L.J.; Gasper, J.R.; Brown, C.D.: Direct and indirect health and safety impacts of electrical generation options. IAEA-SM-254/24 (1981)
18. Devaney, J.J.; Pendergrass, J.H.: The expected environmental consequences and hazards of laser-fusion electric generating stations. In: International conference on health effects of energy production. AECL-6958 (1979)
19. Manthey, Ch. (Ed.): Energy Technology Data Handbook Vol. I – Conversion Technologies. Jül-Spez-70 (Vol. I) (1980)
20. Die Entwicklungsmöglichkeiten der Energiewirtschaft in der Bundesrepublik Deutschland – Untersuchung mit Hilfe eines dynamischen Simulationsmodells. Jül-Spez-1 / Bd. I u. II (1977)
21. Environmental effects of electricity generation. OECD, Paris 1985
22. Energie und Umwelt. Sondergutachten des Rats von Sachverständigen für Umweltfragen. Stuttgart und Mainz (1981)
23. Myers, D.K., et al.: Carcinogenic potential of various energy sources. IAEA-SM-254/02 (1981)
24. Jacobi, W.; Schmier, H.; Schwibach, J.: Comparison of radiation exposure from coal-fired and nuclear power plants in the Federal Republic of Germany. IAEA-SM-254/17 (1981)
25. Matthies, M.; Paretzke, H.G.: The assessment of long-term effects of CO_2 and C-14 due to various energy scenarios. IAEA-SM-254/17 (1981)
26. Bericht über Ursachen und Verhinderungen von Wald-, Gewässer- und Bautenschäden in der Bundesrepublik Deutschland. Der Bundesminister des Inneren 1984

27. Acid rain and transported air pollutants: Implications for public policy. Washington, D.C.: U.S. Congress, Office of Technology Assessment: OTA-0-204 (June 1984)
28. Deutsche Risikostudie Kernkraftwerke – Eine Untersuchung zu dem durch Störfälle in Kernkraftwerken verursachten Risiko. Köln 1979 (English translation: German risk study – main report. A study of the risk due to accidents in nuclear power plants. EPRI NP-1804-SR, April 1981)
29. Darvas, J., et al.: Kontrollierte Kernfusion: eine Alternative der künftigen langfristigen Energieversorgung. In: Münch, E. (Hrsg.), Tatsachen über Kernenergie. Essen 1980
30. Coppola, A.; Hall, R.E.: A risk comparison. NUREG/CR-1916 (1981)
31. Nuclear and non-nuclear risk. An exercise in comparability. EUR 6417 (1980)
32. Bolton, R.S., et al.: Geothermal energy technology. In: Considine, D.M. (Ed.): Energy Technology Handbook. New York 1977
33. Laser, M.: Die Wiederaufarbeitung und Behandlung radioaktiver Abfälle. In: Münch, E. (Hrsg.): Tatsachen über Kernenergie. Essen 1980
34. Albrecht, E.: Die Tieflagerung radioaktiver Abfälle in Salzformationen der Bundesrepublik Deutschland. In: Münch, E. (Hrsg.): Tatsachen über Kernenergie. Essen 1980
35. Falbo, C.: Finite Mathematics applied. Belmont (California) 1977
36. Hauptverband der gewerblichen Berufsgenossenschaften e.V., Hrsg.: Arbeitsunfallstatistik für die Praxis, Bonn 1982
37. Wilson, R.; Chase, W.J.: Problems in the intercomparison of risks of industries and technologies. In: International conference on health effects of energy production. AECL-6958 (1979)
38. Holdren, J.P., et al.: Risks of renewable energy sources: A critique of the Inhaber Report. Report ERG 79-3.Energy and Resources Group, University of California, Berkeley 1979
39. Inhaber, H.: Risks and consequences in energy production. In: International conference on health effects of energy production. AECL-6958 (1979)
40. Whipple, C.: Energy production risks: What perspectives should be taken? Risk Analysis 1 (1981) 29-35
41. Inhaber, H.: Whipple's treatment of energy production risks. Risk Analysis 1 (1981) 37-39
42. Holdren, J.P.: Risks of energy options. Risk Analysis 1 (1981) 41-42
43. Whipple, C.: A reply to comments on energy production risks. Risk Analysis 1 (1981) 43-45
44. Hauptmanns, U.; Sastre, H.: Optimization study for the Spanish electric power-grid. Universidad de Oviedo. Oviedo, Spain 1980
45. Hauptmanns, U.: Risk assessment for interconnected electrical power grids. Risk Analysis 4 (1984) 97-101
46. Auswirkungen des Baus und des Betriebs eines Steinkohlekraftwerks auf Produktion und Erwerbstätigenzahl – Ergebnisse einer Input-Output-Analyse – Wochenbericht des DIW 48/76, 443-448, Berlin 1976
47. Auswirkungen des Baus eines Kernkraftwerks auf Produktion und Erwerbstätigenzahl – Ergebnisse einer Input-Output-Analyse. Wochenbericht des DIW Nr. 26-27, Berlin 1976
48. Deutsches Institut für Wirtschaftsforschung (private communication)
49. Graham, J.D.; Vaupel, J.W.: Value of a life: What difference does it make? Risk Analysis 1 (1981) 89-95
50. Rowe, W.D.: A guidebook for effective analysis and presentation of risks and benefits in alternative energy systems. EUR 11474 EN (1988)
51. House, P.W. et al.: Comparative energy technology alternatives from an environmental perspective. DOE/EV-0109 (1981)

6 Appraisal of the Safety of Nuclear Power Plants

In the preceding chapters the state of the methods and the results of assessing the risks of plants of the nuclear and process industries as well as of non-nuclear energy production are described.

Accidents are possible even in technical installations commonly considered as safe. Absolute safety cannot be achieved, just as there is no absolute safety in other areas of life.

Hazardous industrial plants are subject to official authorization and supervision. The responsible administration has to judge to what extent provisions against damage have to be made during the erection and operation of a proposed industrial plant. Furthermore, it must be decided whether or not the associated risks are tolerable.

Probabilistic analyses allow assessment of the risks of a plant. The effectiveness of the provisions against damage can be quantified. The preceding chapters have shown that in performing such analyses the design and the operational modes of the plants are systematically examined and modelled. In this way experience from plant operation and advances in safety technology may be used to improve plant safety. Safety margins may be quantified and possible weaknesses detected. Furthermore, it may be judged whether design changes requiring less investment are possible without detriment to safety.

In the following the prevailing practice of appraisal of the safety of nuclear power plants is described and areas are identified where procedures need further development. Fundamental concepts for complementing the safety assessment presently in use by probabilistic methods are developed. The treatment places emphasis on the situation in the Federal Republic of Germany.

6.1 Principles of Technical Safety

In all areas of technology there exist laws, rules, guidelines and regulations, as for example references [1 – 5] for nuclear installations in the FRG. They serve to ensure the proper design and functional safety of equipment, means of transportation or industrial installations and to protect man and the environment from harm caused by normal operation and failures. The major part of these procedures were derived from experience with malfunctions, failures, or accidents. Nevertheless, some of them were established anticipating possible failures or accidents without having experienced them.

Complex technical systems are made up of a large number of components, some of which contribute to system safety by their mere presence, i.e. passively (e.g. containments), and others actively (e.g. pumps). Disturbances or accidents result from component failures, which occur for the following reasons:

1. Passive components are designed such that their strength is greater than the expected loads. Since neither the strength nor the loads in real systems may be accurately determined in advance, uncertainties are compensated by overdimensioning. Its degree is described by the safety factor which indicates the relation between the strength of material and the load. Despite overdimensioning, a loss of integrity and hence a failure cannot be excluded. There remains a probability, however small, that the load exceeds the strength.
2. Active components are known to fail with a certain probability. Generally there are different modes of failure. For example, a valve may be stuck and therefore may not open or close again after opening. A motor may fail because of problems with contacts or a faulty controller.

Component performance is quantified by indicating unreliability or unavailability (cf. Sect. 2.1.4.1). Apart from improving the quality of components by a suitable choice of materials and quality assurance during their production, the reliability or availability of technical systems can be increased by redundant and diverse design, spatial segregation of redundant components, quality assurance in production and construction, and an adequate strategy and execution of maintenance.

Systems will only function if their components withstand the loads, fulfil their function and interact as intended. Often systems are designed so that conditions which originate from failures of important components can still be controlled. In the case of hazardous installations, this feature is normally demonstrated in the licensing procedure.

With nuclear power plants those plant conditions that still have to be controlled after the failure of important functions are called design basis accidents. They are usually chosen such that they imply the maximum possible loads and functional requirements on the systems of the plant. The design requirements for safety systems are derived from these design basis accidents. The safety factors to be used and the component reliabilities (quality standards) are generally specified in safety rules and guidelines. The same is true for the functional requirements on the safety systems which are needed for controlling accidents. Calculations for the design of the safety systems are carried out using the deterministic initial and boundary conditions laid down in these rules.

Technical systems which are designed and licensed according to such criteria are accepted as 'safe', although uncontrolled accidents remain possible. However, their expected frequencies of occurrence are generally small.

The outlined historically grown safety philosophy is still acceptable today, if

- a technology has matured long enough to allow sufficient experience (including negative) to be gathered, or
- technical systems can be put on trial before their commercial use, or
- the potential consequences of accidents are relatively limited.

For nuclear power stations as a whole none of these three conditions applies; however, the first two conditions are satisfied by most of their components and systems.

6.2 Outline of Legal Foundations

On the basis of the Constitution of the Federal Republic of Germany and the Atomic Energy Act [1] nuclear power plants are generally licensable, even if this implies tolerating certain risks. In order to guarantee ample protection from the hazards of nuclear energy the Atomic Energy Act subjects the use of nuclear fuel and other radioactive materials to control and supervision by government authorities. In contrast to the licensing of other technical installations which must be licensed if they satisfy all requirements nuclear licensing requirements are formulated as rules of prohibition with the proviso of permission. Accordingly, a nuclear power plant can be licensed if the necessary protection from damage which might be caused by its erection and operation is ensured according to the state of science and technology.

This notion of necessary protection must be made concrete and interpreted. The objective is the protection from danger which is assumed to exist whenever the uncontrolled progression of events may lead to damage with a sufficiently large probability. In this context, sufficiently large probability does not imply a fixed numerical value. The degree of uncertainty of occurrence of the damage, i.e. its probability of occurrence, must be judged with regard to the need for protection of the endangered rights and property and to the expected magnitude of damage. If the damage can be very large, already the remote probability of its occurrence would be sufficient to assume danger. In view of the possible catastrophic consequences of nuclear accidents the application of the corresponding legislation assumes that danger already exists for accidents with very low probabilities of occurrence. Events even less likely to occur than those accounted for to ensure adequate protection of the individual have to be taken into consideration to protect the public [6].

The assessment and appraisal of risk is the responsibility of the executive. The possibility of damage must be accounted for even if the sole reason for not excluding it is insufficient knowledge on whether it will be produced or not. Danger may then merely be suspected but not assumed; there is just a potential for concern. In appraising safety, uncertainties of judgement on probabilities have to be accounted for by sufficiently conservative assumptions.

The demonstration of adequate protection of the individual and the public from danger is an indispensable requirement in licensing. In particular, a licence must be denied if the plant may cause damage which affects the rights or property of the individual, as protected by the Constitution. The individual whose rights or property are affected is entitled to an actionable claim to sufficient protection. The obligations of government authorities to protect the public go even beyond the elimination of danger to the rights and property of the individual. However, the individual is not legally entitled to claim this type of protection from the authorities.

The legislator has granted margins of judgement and valuation to the authorities in order to:

− account for the novelty of nuclear technology and the uncertainty in its appraisal;
− provide the possibility of accounting for foreseeable future developments;
− comply with the obligation of planning in agreement with the law, i.e. to guarantee comprehensive and balanced deliberations which account for all important factors.

The authorities are obliged to exercise their judgement dutifully, to make use of all sources of scientific and technical knowledge, and to decide without arbitrariness. A claimant whose rights may be affected is entitled to faultless judgement by the authorities.

Dutiful judgement by the authorities implies that

− relevant circumstances are correctly ascertained;
− proper deliberations are made;
− the principle of equality is respected;
− prescribed procedures are observed.

With the Atomic Energy Act the legislator has provided a legal instrument which obliges the executive to act according to the principle of 'best possible protection from danger and precaution against risk'.

The distinction between the two areas of provision, protection from danger and precaution against risk, can be interpreted as follows:

1. 'Protection from danger' comprises all provisions which are required for the protection from danger, in particular to the rights and property of the individual, regardless of expense and technical viability. The potentially affected individual is entitled to an actionable claim.

   'Best possible protection from danger' describes the obligation of the authorities to examine the necessity of precautions against damage and their adequacy on the basis of the best information available, namely according to the state of science and technology.

2. 'Precaution against risk' goes beyond the protection of the rights and property of the individual from danger and comprises provisions for mitigating impacts or reducing the frequencies of occurrence and consequences of events capable of causing damage. Precaution against risk may be demanded in the process of appraisal or, in many cases, because of legal ordinances derived from [1]. The individual is not legally entitled to claim specific precautions against risk.

'Best possible precaution against risk' describes the execution of dutiful judgement which accounts for all important objectives. For example, differing requirements for protecting the population, the environment, and employees and for achieving operational safety have to be balanced against the benefit to the economy and the prospects of development of nuclear technology. Conflicts of objectives have to be resolved as far as possible. The principle of adequacy of cost and benefit has to be observed. The dutiful execution of judgement is subject to judicial control in the case of disputes.

"Only the continuous adaptation of the circumstances relevant for risk evaluation to the most recent state of knowledge can satisfy the principle of best possible protection from danger and precaution against risk" [7].

The method of probabilistic safety analysis reflects the current state of science and technology. Consequently, probabilistic safety assessment has to be used for safety relevant decisions as far as appropriate for the protection demanded in the Atomic Energy Act. Decisions on safety which disregard the use of probabilistic safety analysis without sufficient justification would be at fault.

With regard to the applicable standards, the following problems of delineation arise in the execution of the law:

I.   Delineation between the required protection from danger and the wider precaution against risk.
II. Delineation of measures of protection of the individual as against those precautions for protecting the public which cannot be legally claimed by the individual.

Table 6.1 shows the four areas of provision to be distinguished. The delineations between the different areas are important for the practice of appraising technical safety. They determine:

- the extent to which safety has to be demonstrated,
- the level of probabilities of occurrence of events for which provisions have to be demonstrated,
- the type and magnitude of consequences which may be assumed as tolerable in the planning and implementation of precautions against certain events, and
- the extent to which compensation between costs and benefits of provisions is permitted.

Table 6.1. Legal delineation of the different areas of provision in licensing nuclear power plants according to the types of decision and of rights to be protected

Type of decision of the authority	Bound decision: compulsory prerequisite for licensing	Judgemental decision based on the margin of decision
Rights to be protected:		
Protection of the rights of the individual	Provisions which are necessary regardless of cost and technical viability for protecting the rights and property of the individual from danger	Provisions for mitigating risks for the life, health, and property of the individual, which can be implemented without detriment to the other objectives of the legislation for nuclear installations
Protection of the public	Provisions beyond the protection of the individual required because of the potential magnitude of the consequences and the uncertainties of assessment	Provisions implemented observing the principle of adequacy and having compared the different objectives of protecting all rights contemplated in the Atomic Energy Act

The delineation problems, however, cannot be resolved by using abstract notions or formal criteria, but only in the course of appraising technical safety. Therefore, the legislator has deliberately conferred to the executive the task of judging the protection from danger and the precautions against risk necessary on the basis of the state of science and technology.

6.3   Procedure of Safety Evaluation

6.3.1   Foundations

Before the first licensing of a nuclear power plant in the FRG there was little experience of nuclear safety. All those responsible were conscious of the fact that the licensing of nuclear power stations meant the authorization of new and complex technical installations with a large hazard potential. Therefore the objective was to develop a feasible procedure to demonstrate the licensability of nuclear power stations under the requirements of the Atomic Energy Act [1]. This was possible with the multi-barrier system (cf. Fig. 3.1), also called defence in depth, combined with the concept of four safety levels for protecting these barriers, as shown in Fig. 6.1.

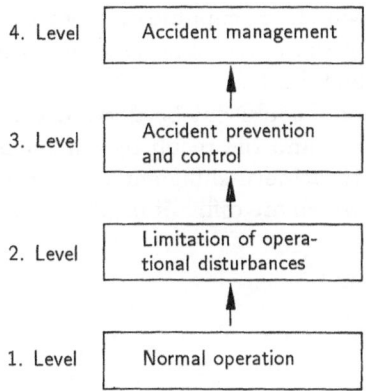

4. Level | Accident management

3. Level | Accident prevention and control

2. Level | Limitation of operational disturbances

1. Level | Normal operation

Fig. 6.1. Concept of the four safety levels – preventive and mitigating measures for the protection from damage

Having established the feasibility of the design for safety and consolidation of the underlying safety concept, the present concern is to benefit from scientific and technological progress for continual improvement of the safety of nuclear power plants.

The concept of the safety evaluation of light water reactors is laid down in the safety criteria for nuclear power plants of the BMI (Federal Minister of the Interior) [3]. Accordingly, a nuclear power station has to be designed by measures of the first safety level so that operation without disturbances and adverse impact on the environment is possible. For measures of the second safety level all events have to be taken into account by the design which are expected to occur during the period of operation of the plant, particularly operational disturbances. The integrity of the barriers enclosing the radioactive materials has to be guaranteed to such an extent that these events do not lead to exposures of individuals in the surroundings of the plant beyond the dose limits of the Radiation Protection Ordinance, i.e. $0.3/0.9 \, \text{mSv a}^{-1}$ ($30/90 \, \text{mrem a}^{-1}$; $0.3 \, \text{mSv a}^{-1}$ effective dose and $1.8 \, \text{mSv a}^{-1}$ partial body dose to bone surfaces and skin via the water and atmospheric exposure paths, and $0.9 \, \text{mSv a}^{-1}$ partial body dose to all other organs from all food chains). The design for normal operation is not sufficient to demonstrate that the protection from damage required by the Atomic Energy Act is guaranteed. Important deviations from this operational regime, called 'incidents' in the German legislation, are improbable; however, they cannot be excluded. Due to their large hazard potential they may cause considerable damage, unless adequate precautions are taken. For this reason, additional protective measures are required on the third safety level, that of accident control. The objective of these measures is to guarantee the effectiveness of the barrier functions as far as possible. In particular, the following safety functions must be ensured:

– control of reactivity, i.e. of the energy release from the reactor core;
– cooling of the reactor core;
– availability of a heat sink;
– safe containment.

For this purpose safety systems are required which are capable of protecting the plant from inadmissible loads and of keeping the impact on personnel, the plant, and the environment within pre-established limits.

Technical specifications for these safety systems could not be derived from experience. The progression of potential accidents and the resulting loads can often only be analysed theoretically. In order to achieve sufficient protection from damage, plant conditions must be avoided which are difficult or impossible to control, in particular excessive core heat-up.

6.3.2 Present Practice

The design against accidents is based on a limited number of design basis accidents, as already explained in Sect. 6.1. A licence can only be granted if it is demonstrated that, taking into account the design principles and predetermined boundary and initial conditions, the plant can cope with these design basis accidents. The effectiveness of the barriers for containing the radioactive materials must be preserved such that the releases resulting from an accident do not lead to exposures above the admissible values of 0.05/0.15 Sv (0.05 Sv effective dose, 0.15 Sv thyroid dose, 0.3 Sv bone surface dose) in the vicinity of the nuclear power plant [2].

By measures on the first three levels of safety, significant damage to the first two barriers (the reactor core) is prevented for all events which cannot be excluded during the lifetime of a large number of reactors.

Finally, on the fourth level measures are taken to reduce the frequencies or consequences of any event sequences for which the plant is not designed and does not have to be designed because of their extremely low probability of occurrence. It would contradict the principle of adequacy to require the same procedures and safety targets as for the design basis accidents for event sequences which can only occur if safety measures have failed.

For planning and appraising measures on the fourth level of safety two objectives have to be distinguished:

1. Reduction of the expected frequencies of occurrence of events which could threaten the integrity of the barriers.
2. Mitigation of damage by strengthening already existing barriers or adding further barriers.

The assessment of the safety of nuclear power plants is performed primarily on the basis of the design requirements laid down in the relevant legislation, rules and regulations. This is especially true for the first three levels of safety. This procedure is called 'deterministic'. The design basis accidents are not selected according to their expected frequencies of occurrence, i.e. probabilistically. The chief objective is to represent as many deviations from normal operation as possible by a minimum number of design basis accidents. Hence, the deterministic safety evaluation is not a risk assessment. The reliability and effectiveness of operational and safety systems are not judged on the basis of the probability of their failure. Nevertheless, the single failure criterion and the requirements of redundancy and diversity imply probabilistic considerations, which, however, remain unquantified.

The deterministic procedure with its restriction to a limited number of decisive safety requirements represents a pragmatic and practicable basis for licensing which

- shows the executive how the licensability of nuclear power plants can be demonstrated in compliance with the Atomic Energy Act and
- sets sufficiently clear goals to the engineer which must be met by the design of a nuclear power plant.

This procedure is supported by supervision, evaluation of operating experience, and optimization using the results of reactor safety research.

According to the practice of judging the safety of a nuclear power plant deterministically the protection of the individual from damage is mainly achieved through measures of the first three levels of safety, which guarantee:

- compliance with the $0.3/0.9\,\mathrm{mSv\,a^{-1}}$ radiation exposure for the most exposed individual in the vicinity of the plant during normal operation (first and second levels);
- that the radiation exposure admissible for accidents in the vicinity of the plant can virtually not be exceeded because of structural and technical measures (third level) and additional provisions.

The accidents which are to be covered by the design basis accidents in order to guarantee the required protection from damage have been selected primarily on the ground of their hazard potential. The radiation safety goal on which the planning of protective measures against design basis accidents has to be based refers to the most severely affected individual in the vicinity of the plant. Regarding the limitation of consequences from extremely unlikely accidents this requires the demonstration of provisions which reach beyond the needs of protection of the public [2]. Therefore, this procedure guarantees not only the required protection of the individual and the public from danger but also contributes to precaution against risk. All provisions beyond this serve to reduce the risk for all relevant rights and property which have to be protected, in particular for the individual in the vicinity of the plant.

For pressurized water reactors of modern design the delineation between the different areas of provision has been outlined in the Incident Guidelines of 1983 [8]. These Guidelines contain a complete list of the design basis accidents relevant to the third level of safety. For every design basis accident they fix the boundary conditions for the analyses and the type of proof to be furnished. With the safety requirements for normal operation – first and second levels of safety – and the relevant rules the executive has determined that further measures for the protection of the individual are considered unnecessary. The Incident Guidelines make further clarifications in order to

- avoid an inadequately high effort for safety and its demonstration
  (Radiation impact has to be calculated only for a limited number of design basis accidents which are considered to be representative of all design basis accidents. Other design basis accidents have to be analysed only with regard to the design of the safety functions or of the components. For certain design basis accidents it is sufficient to show that specific provisions are made. Event sequence analyses are not required in such cases.);
- ensure sufficient protection from extremely unlikely accidents which have potentially severe consequences
  (Besides the area of provisions covered by the list of design basis accidents, other extremely unlikely sequences are conceivable for which provisions have to be considered because their consequences can be extremely severe. The Guidelines name, for example, an aeroplane crashing on the reactor or the failure of the reactor scram system during operational transients.).

In these circumstances the plant must be capable of avoiding or limiting damage. The relevant rules contain certain requirements – for example, concerning the pressure resistance of the primary circuit – or technical specifications for certain event sequences like load-time diagrams [9]. The Incident Guidelines state that such events are not design basis accidents because of their low risk and therefore not the subject of the Guidelines.

The outlined simplification of the proof of compliance implies a sufficiently concrete plant design with well-defined features, as for example the modern pressurized water reactors of Siemens-Kraftwerkunion (KWU).

This procedure provides a proven and consolidated framework for the safety valuation of nuclear power plants, because:

- numerous plants in operation or under construction, for which compliance with licensing requirements has been demonstrated, have set technical standards for future nuclear power plants, and
- the rules furnish a broad basis for demonstrating the licensability of nuclear power plants.

This framework represents the starting point for a more differentiated appraisal of the technical and safety design of nuclear power plants.

6.3.3   Further Development

The Incident Guidelines are based on experience with safety analyses and their evaluation, on insights from many years of operation of pressurized water reactors, and on research results. They reflect the practice of licensing in recent years as do the technical rules below the legally binding provisions which are not opposed to modifications of the prevailing procedures. They allow safety systems and operational procedures to differ from present practice if the protection demanded by the Atomic Energy Act is guaranteed. Proof that modifications or new designs provide sufficient or even better provisions is the responsibility of those who request such changes. The level of safety achieved in modern nuclear power plants as documented in the technical rules is the yardstick for decisions on requirements, suitability or adequacy as well as balance. It has to be examined which area of provision – protection from danger or precaution against risk – is affected. This then determines the technical and legal standards to be applied and the proofs of compliance to be furnished.

The deterministic concept sets rules for the demonstration of the required protection of rights and property of the individual. Yet, it meets limitations, in particular with regard to the optimization of plant operation and protection from damage. In the formal application of deterministic criteria, frequent event sequences are not distinguished from rare ones. Reliable components or systems are not treated any differently from less reliable ones. This can lead to imbalance and even be detrimental to safety or can produce conflict between different safety objectives. Because of the separate treatment of different design areas, which is characteristic of deterministic analyses, safety devices may be excessively optimized with regard to specific aspects. Interactions, counteracting influences or intermeshing may not be sufficiently accounted for. Furthermore, the deterministic concept of design basis accidents has the limitation of assuming that other accidents pose lesser requirements on safety systems and will therefore be coped with, without this being always demonstrated. On the other hand, it implies that event sequences conceivable beyond the predetermined design limits do not have to be considered in the licensing process. An important tool for a more differentiating safety valuation are probabilistic methods. Their utilization corresponds to the legal obligation to further the protection from danger and precaution against risk according to the possibilities and necessities derived from the state of science and technology.

The practice of licensing implies that the proof of absolute exclusion of damage is not possible and is also not required by the law. At the time of licensing no evidence must be available which could doubt that the required protection from danger has not been achieved. The authorities must be aware of the possibility that their decisions could have been made on the 'basis of a not yet disproved scientific error'. The authorities must therefore re-examine prevailing assumptions in view of new insight and take care that suitable improved measures of protection are implemented. The legislator also allows retrogressive injunctions inasmuch as they are required to achieve protection according to the Atomic Energy Act and regulates whether or not the responsible authority is obliged to compensate in such cases. Such legal rules not only provide opportunities but also establish the obligation of examinations or actions by the authorities. These may be supported efficiently by using probabilistic methods so that the safety of nuclear power plants can be improved by the joint application of deterministic and probabilistic procedures.

6.4 Probabilistic Characterization of the Present Level of Safety

6.4.1 Protection of the Individual

Based on results of probabilistic safety analyses, the level of protection of the individual from danger, which is considered sufficient in view of the requirements of the law, may be summarized as follows:

- During specified normal operation the dose limit of $0.3/0.9 \, \text{mSv} \, \text{a}^{-1}$ for the individual with the highest exposure in the vicinity of the plant must not be exceeded. Specified normal operation comprises all plant conditions (normal operation, operational disturbances, maintenance etc.) which are expected to occur during the lifetime of the plant, i.e. with an expected frequency of occurrence greater than 10^{-2} to 10^{-3} per year.
- The sum of the expected frequencies of occurrence of all accidents by which the admissible values for radiation exposure could be exceeded is smaller than 10^{-4} per plant and year; the expected frequencies of occurrence of individual representative event sequences lie below 10^{-5} per plant and year.
- In addition, the plant has certain consequence-mitigating properties for events which are less probable.

6.4.2 Precaution Against Risk

Precautions against risk which go beyond the safety level just described may especially be beneficial for individuals in the vicinity of the plant. Figure 6.2 serves as an explanation.

Accordingly, four regions of precautions against the risk from radiation exposure of individuals in the surroundings of the plant may be distinguished. They correspond to the four safety levels. The areas are designated in Fig. 6.2 with the letters A_1/A_2, B, and C. They are explained in Table 6.2.

Precautions against risk affecting the areas A_1/A_2 have to be taken according to the relevant legislation which requires that radiation exposure has to be kept as low as possible, even below the admissible limits. In particular, deviations from normal operation have to be avoided.

Precautions against risk inscribed in the areas B and C derive from the objective to limit safety relevant deviations from normal operation as early and with as few consequences as possible for people, plant and environment. This can be achieved for example by:

- increasing the reliability or the availability of systems for controlling possible deviations from normal operation and by limiting the consequences of such occurrences;

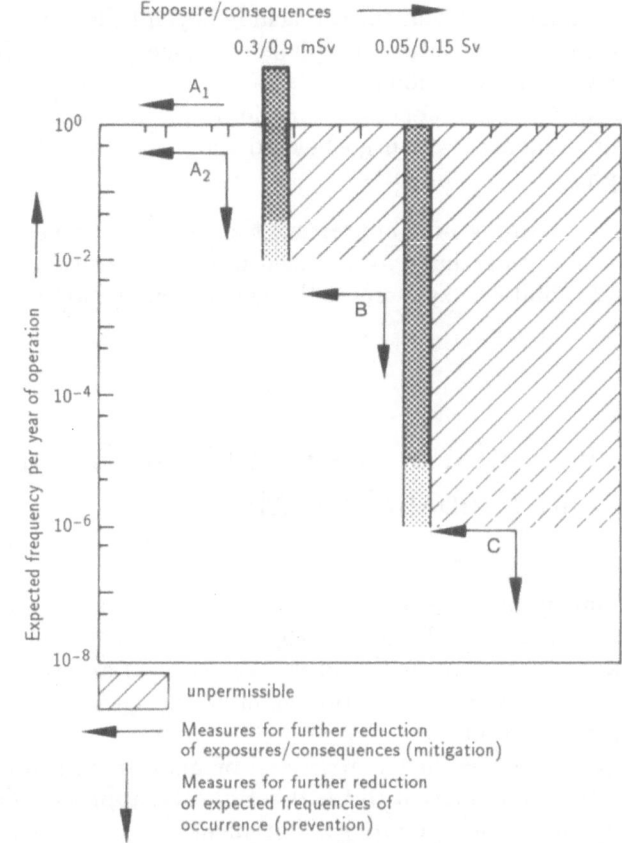

Fig. 6.2. Radiation-protection goals and risk-reducing effects of additional measures in the different areas of provision

Table 6.2. Objectives of the precautions against risk

A_1: Avoidance of unnecessary radiation exposure, reduction of the emission of radioactive substances as a precaution against conceivable consequences from synergistic or accumulative effects; minimization of radiation exposure.

A_2: Limitation of the frequency of occurrence and of the consequences of operational disturbances in order to
   - reduce the emission per event,
   - reduce the occupational radiation exposure,
   - protect systems and components,
   - avoid plant conditions which can progress to an accident.

B: Limitation of frequencies of occurrence and of consequences of accidents
   - as for A_1 and A_2,
   - avoidance of plant internal contamination, of radioactive waste, and repairs,
   - avoidance of event sequences going beyond the design basis.

C: Prevention of severe accidents and mitigation of accident consequences
   - avoidance of off-plant contamination and radiation exposure of people and environment,
   - avoidance of long-term impacts by clearing debris and managing wastes,
   - avoidance of burdens for the economy and of losses including that of an energy source.

- controlling event sequences reaching beyond the design basis, for which it is still possible to return the plant to a safe condition (cold subcritical with intact barrier functions);
- mitigating consequences or providing for the elimination of damage from event sequences which go beyond those mentioned in the preceding paragraph.

When planning and appraising the precautions against risks all types of impact or damage have to be taken into consideration, for example the occupational radiation exposure or the generation of radioactive waste.

6.5   Potential of the Combined Use of Deterministic and Probabilistic Methods

In regulatory practice based on the deterministic approach it is assumed that the design basis accidents (cf. [10]) are controlled. The resulting accident sequences are represented by the success paths of an event tree, which are characterized by the proper functioning of all the systems required for coping with the initiating event.

Although not explicitly expressed or quantified, a probabilistic judgement has always been inherent to the deterministic approach. On the different levels of the hierarchy of regulatory requirements, safety rules and criteria often contain qualitative probabilistic concepts concerning frequencies and consequences of events such as: sufficiently reliable; negligibly small; remote; highly improbable; no unreasonable risk; as low as reasonably achievable (ALARA-principle).

With matured methods and improved data bases the probabilistic approach is being increasingly accepted. It is not only an additional tool for assessing the safety of a nuclear facility but also provides an information base which is applicable to a wide variety of issues and decisions. Therefore deterministic and probabilistic approaches are now jointly being used in evaluating and improving nuclear safety.

Probabilistic analysis allows quantification of the probabilities of the functioning of the systems required for controlling the initiating event and hence calculation of the probability for coping with the accident. This is an extension of the deterministic concept, which assumes functioning of all these systems and hence an availability of 1. This extension implies that the success paths are complemented by a set of conceivable event sequences which result from the failure of one or several safety systems. If the systems are well designed, the event sequence triggered by an initiating event progresses with a probability close to 1 along the success paths. All other event sequences are improbable. This probabilistic approach complements the deterministic approach but cannot replace it. It provides insight into the adequacy of the existing safety systems for initiating events which lead on to others than the design basis accidents. In this way it can be judged whether the design basis accidents have been properly selected. The need to cope with different types of accidents may lead to contradictory requirements on the safety system. The probabilistic approach then enables a more balanced concept to be developed.

Taking, for example, the results of the DRS-A (cf. Sect. 3.2.4.1) which showed a dominant contribution to the risk of a small leak in the primary coolant circuit, it might be asked whether the design basis accident 'large break' should be replaced by a design basis accident 'small leak'. A renewed probabilistic analysis would then have to show whether the design changes resulting from this do not increase the contributions from other initiating events to the core-melt frequency. In this context it should be noted that the maximum loads on the primary coolant circuit and the containment, especially differential pressure loads on their internals, are caused by the pressure wave after a large break. Therefore this design basis accident should be retained when dimensioning the primary circuit and the containment.

This example indicates how the probabilistic approach can be used for the selection of appropriate design basis accidents.

Table 6.3 shows how questions to be asked in this context may be related to the results from probabilistic analyses.

As probabilistic analyses also address event sequences beyond the design basis they may serve as well to:

- indicate the level of nuclear power plant safety and to quantify safety margins beyond the design limits as a reference for future safety decisions;
- concentrate resources on preventive and mitigating features most important to risk and to show how to improve control of the initiation and progression of events which may lead to severe accidents;
- develop strategies for coping with accidents beyond the current design basis; for example, to identify possibilities of diagnosing the more probable severe accident sequences, to provide information and guidance to the operators to deal with such accidents and to supply information for developing accident-management procedures, all of which pertains to the fourth level of safety (see Fig. 6.1).

When making use of the results of probabilistic analyses their limitations should be borne in mind.

Systems analysis for internal initiating events has reached a high level of quality, but is less satisfactory in the case of external events.

The data base is fairly good for events which can frequently be observed in the current population of reactors, but is still poor for rare events.

The modelling of common cause failures (cf. Sect. 2.1.4.7) and of human error (cf. Sect. 2.1.4.8) and estimates of source terms (cf. Sect. 3.1.1.4) have improved during the past years, but are still affected by considerable uncertainties.

Table 6.3. Probabilistic methods as decision aids

Question	Answered by:
Quality of the safety systems?	Unavailabilities on demand and unreliabilities of systems.
Quality of the design of the plant?	Sum of the expected frequencies of core melt. Modes, times and probabilities of containment failure. Release categories.
Overall balance of design of the plant?	Differences between the expected frequencies of occurrence of the different accident sequences.
Dominant contributions to risk?	Investigation of 1. Events which figure in many minimal cut sets. 2. Events which figure in minimal cut sets consisting of few elements in order to identify components and functions which would change the core-melt frequency or the risk substantially if their failure probabilities were modified. 3. Probabilities of the minimal cut sets of dominating sequences.
Effective methods for mitigating accident consequences?	1. Evaluation of the containment fault and event trees. 2. Valuation of the results of the accident-consequence calculations (seeking shelter, time history and extent of evacuation measures).
Magnitude of the risk of different types of damage?	CCDFs and their subjective confidence regions.
Research and development work contributing most to risk reduction?	Improvement of the knowledge on phenomena which lead to substantial uncertainties in calculating the risk-dominant event sequences.

Completeness of the analysis cannot be proved. But this should not be a major drawback when examining design weaknesses and dominant accident sequences.

It is important to note that in probabilistic analyses the uncertainties of the results are quantified. For deterministic calculations which are also affected by uncertainties these are normally not indicated.

6.6   Probabilistic Objectives for Safety

6.6.1   Introduction

In applying probabilistic methods for safety evaluation reliability analyses of the safety systems, the quantification of specific safety parameters or a PRA or PSA of different levels and degrees of profundity may be required. When judging probabilistic parameters relative criteria such as the degree of balance may be used or absolute ones such as quantitative safety goals. Accordingly, probabilistic analyses then have to be carried out to achieve certain objectives, for example the determination of:

– system reliabilities and availabilities,
– frequencies of plant damage states, core degradation or core melt,
– frequencies and magnitudes of releases of radioactive substances into the atmosphere,
– frequencies and magnitudes of radiation exposure in the vicinity of the site,
– health risks for individuals and the population in the vicinity of the site,
– frequencies and consequences from external events and fires.

In the following a survey of quantitative criteria existing in different countries classified according to these objectives is given.

6.6.2   System Reliability and Availability

Objectives concerning operational and safety systems exist, for example, in Canada, Finland and the United Kingdom.

In *Canada* [11] the licensing authority has set bounds for the maximum permissible failure frequencies and unavailabilities for plant systems. They amount to 0.3 per reactor year for all operational systems and 10^{-3} per demand for specific safety systems and containment isolation. Hence, the upper limit for the simultaneous failure of an operational and a safety system is $3 \cdot 10^{-4}$ per year.

A value of approximately $10^{-7}\,\mathrm{a}^{-1}$ is used as cut-off criterion for risk assessment. This corresponds to the simultaneous failure of two independent safety systems after demand due to the failure of an operational system, i.e.

$$h = 0.3\,\mathrm{a}^{-1} \cdot 10^{-3} \cdot 10^{-3} = 3 \cdot 10^{-7}\,\mathrm{a}^{-1}$$

Accidents with an expected annual frequency of occurrence below this value are not considered.

In *Finland* [12] a design objective of a failure probability on demand of 10^{-4} is used for most safety functions. However, a probability of failure on demand of 10^{-5} for making the reactor subcritical and of $5 \cdot 10^{-3}$ for the isolation of the containment is to be achieved. The requirements must be met with a confidence level of 90%.

In the *United Kingdom* [11] a principle is in use which stipulates that a safety system should usually not have an unavailability greater than 10^{-3} per demand and that the unavailability of a system employing identical redundancies is to be limited by values between 10^{-3} and 10^{-5} per demand due to possible common cause failures.

6.6.3   Dangerous Plant Conditions

Dangerous plant conditions contemplated in probabilistic criteria in Italy and the USA are core degradation or core melt.

In *Italy* [12] so far there exist no official probabilistic risk criteria. The Italian Licensing Authority has, however, carried out certain activities for defining such criteria.

In this context a target of 10^{-5} was fixed for the annual frequency of any accident sequence involving core degradation which is assumed to occur when the design basis fuel conditions are exceeded. There is a tendency to lower this value to $10^{-6} \, \mathrm{a}^{-1}$.

In the *USA* several institutions (ACRS, NRC, AIF) and individuals (Bernero, Zebrowski) have in the past proposed bounds for the frequency of core melt, the figure usually being $10^{-4} \, \mathrm{a}^{-1}$. This value was used during the trial period of the safety goals. However, safety goals concerning public health have priority over the core-melt frequency limit. As can be seen from Tables 3.3 and 3.4, the value of $10^{-4} \, \mathrm{a}^{-1}$ is exceeded by some of the plants. The core-melt frequency limit is not included in the present version of the safety goals [13].

6.6.4   Objectives Concerning Radioactive Releases
into the Atmosphere

Objectives concerning radioactive releases into the atmosphere exist, for example, in Canada, France, the UK and the USA.

In *Canada* there is a general objective that the use of nuclear energy should not cause greater risks than the generation of electricity from coal or the operation of other hazardous industries. The requirement is subsequently derived that the frequency of a major accident in a nuclear power station leading to large radioactive releases into the environment should not exceed $3 \cdot 10^{-7}$ per reactor year.

The safety objective adopted in *France* [12] for PWR stipulates that the overall frequency of unacceptable consequences shall not exceed 10^{-6} per year. This implies that unacceptable consequences from individual groups of initiating events should not occur with an annual frequency above 10^{-7}. This threshold must not be exceeded unless it can be proved that the calculation of the relevant frequencies is overly conservative.

The unacceptable consequences mentioned above are not defined by any legislative or regulatory text.

In *Italy* [12] a probabilistic target for limiting the frequency of occurrence of large releases due to catastrophic containment failures has been formulated. Accordingly large releases after a core-melt accident should be prevented with a conditional probability of 0.95. A release is considered as unacceptable if more than 0.1% of the core inventory of I and Cs are released.

In the *United Kingdom* the Design Safety Criteria of the Central Electricity Generating Board (CEGB) [14] provide a target concerning 'uncontrolled releases'. This target requires that the frequency of all such releases should be less than 10^{-6} per year and the frequency of any single release of that kind should be less than 10^{-7} per year. Since these criteria are regarded as extremely restrictive, any failure to comply with them would not automatically rule a reactor system out of consideration.

In the *USA* a safety goal of $10^{-6}\,\mathrm{a}^{-1}$ for a large release of radioactive substances, i.e a release with a potential for causing an offsite early fatality, is presently under discussion [15].

6.6.5 Safety Goals Concerning Radiation Exposure

Safety goals concerning radiation exposure exist, for example, in Argentina, Canada, France, Switzerland, and the United Kingdom.

In *Argentina* [16] the criterion curve of the Comisión de Energía Atómica, shown in Fig. 6.3, is in use. It relates the frequency of accidents to maximum permissible doses of radiation. The utility proposing the construction of a nuclear power station has to prove that all conceivable incidents lie within the area denoted as 'acceptable'.

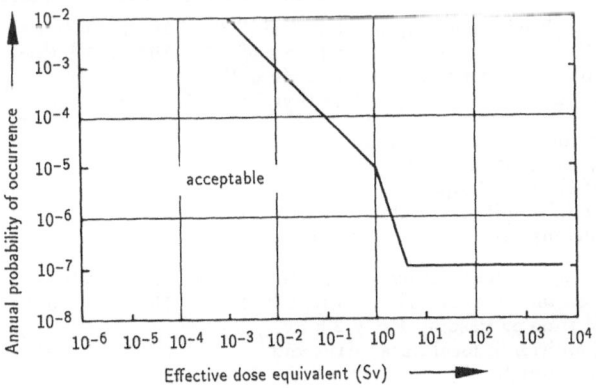

Fig. 6.3. Criterion curve of the Comisión Nacional de Energía Atómica of Argentina

In order to demonstrate this, the event sequences which lead to a radiation exposure of the population must be identified. Using event and fault tree analyses (cf. Sect. 2.1), the corresponding expected frequencies have to be calculated. In doing this 'generally accepted' reliability data are to be used. Exceptions are possible if a different choice can be justified. If reliability data for some types of components are not available, the values to be employed are proposed by the licensing authority. Details of the procedure are regulated in [17].

In *Canada* the criteria for radiation exposure shown in Tables 6.5 and 6.6 are in use.

As mentioned before, there is a cut-off criterion of $10^{-7}\, a^{-1}$. In addition, the operational and reference dose limits of Table 6.6 have to be observed. These values are shown graphically in Fig. 6.4. It should be noted that in the low dose range the corresponding frequencies are chosen such that risk is constant. Higher doses, however, are weighted more strongly.

Table 6.5. Dose-frequency relationship in Canada

Maximum admissible annual frequency	Individual whole body dose in Sv	Individual dose for the thyroid in Sv
10^{-1}	0 ...0.0005	0 ... 0.005
10^{-2}	0.0005...0.005	0.005... 0.05
10^{-3}	0.005 ...0.05	0.05 ... 0.5
10^{-4}	0.05 ...0.1	0.5 ... 1
10^{-5}	0.1 ...0.3	1 ... 3
10^{-6}	0.3 ...1	3 ...10

Table 6.6. Canadian operational and reference dose limits for accident conditions

Situation	Assumed maximum frequency	Meteorology to be used in calculation	Maximum individual dose limits	Maximum total population dose limits
Normal operation		Weighted according to effect, i.e. frequency times dose for unit release	0.005 Sv a^{-1} whole body 0.03 Sv a^{-1} to thyroid	10^2 man-Sv a^{-1} 10^2 Sv a^{-1} thyroid
Serious process failure (single failure)	1 in 3 years	Either worst weather existing at most 10% of time or Pasquill F condition if local data incomplete	Same as above	
Serious process failure plus failure of any safety system (dual failure)	1 in 3000 years	Either worst weather existing at most 10% of time or Pasquill F condition if local data incomplete	0.25 Sv whole body 2.5 Sv thyroid	10^4 man-Sv 10^4 Sv thyroid

The terms 'single failure' and 'dual failure' used in Table 6.6 and Fig. 6.4 are defined as follows:

- single failure: an operational system fails totally,
- dual failure: an operational system fails and at the same time the function of a safety system is impaired or lost.

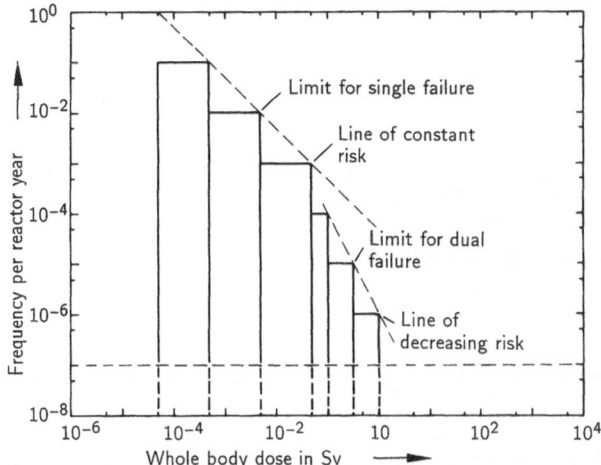

Fig. 6.4. Canadian safety criteria for population doses

In *Italy* [12] the following radiological targets have been defined by the licensing authority:

- 0.1 mSv per year for normal operation and anticipated operational transients;
- 5 mSv for accidents whose annual frequency of occurrence is higher than 10^{-3};
- 100 mSv for accidents whose annual frequency of occurrence is below 10^{-3}.

A further reduction of the dose values is recommended if the plant can be improved at reasonable expense. In particular, a value of 5 mSv for design basis accidents is the goal to be achieved.

In *Switzerland* the limiting curve shown in Fig. 6.5 is used as a design criterion which serves for orientation. It relates the frequency of occurrence of 'incidents' and accidents to maximum whole body doses. The following limits are applied:

- 0.3 mSv for normal operation and operational disturbances, whose annual frequency of occurrence is greater than 10^{-2};
- 1.0 mSv for 'incidents', whose annual frequency of occurrence lies in the range from 10^{-2} to 10^{-4};

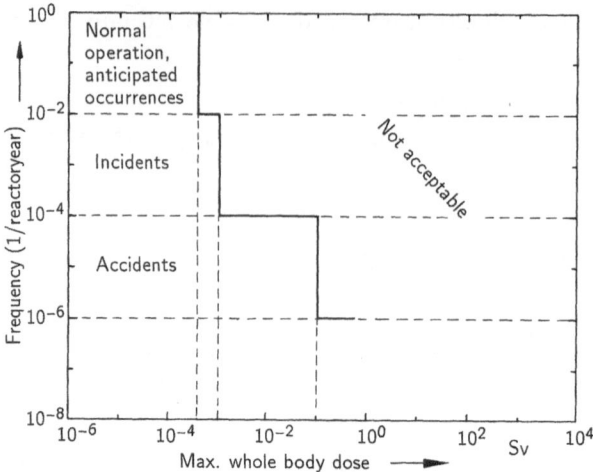

Fig. 6.5. Probabilistic design criterion in Switzerland

- 0.1 Sv for accidents, whose annual frequency of occurrence lies between 10^{-4} and 10^{-6}.

In the *United Kingdom* the utilities themselves are responsible for the safe design and operation of nuclear power plants, for which purpose they developed certain criteria. These have been judged appropriate by the Nuclear Installations Inspectorate (NII). Some of the criteria are formulated as dose limits related to frequencies of occurrence. They are expressed in terms of the emergency reference level (ERL) which, for example in the case of the whole body dose, amounts to 0.1 Sv. In particular we have:

- with an expected annual frequency of 10^{-2} a release between 1/1000 and 1/100 ERL;
- with an expected annual frequency of 10^{-3} a release between 1/100 and 1/10 ERL;
- with an expected annual frequency of 10^{-4} a release of the order of the emergency reference level;
- event sequences leading to magnitudes of damage above the emergency reference levels have to be made as improbable as reasonably practicable.

If it cannot be proved that these requirements are satisfied for an accident sequence so-called effective barriers are demanded by the NII as an additional safety measure. They may be realized by active or passive equipment. Every barrier is expected to have an unavailability of 10^{-4} per demand. Accordingly (see Fig. 6.6):

- with accidents leading to a dose equivalent below the emergency reference level the existence of one effective barrier has to be proved;

– with accidents which are expected to occur with an annual frequency below 10^{-3} to 10^{-4} and cause radiation exposures of the order of magnitude of the emergency reference level one effective barrier has to be provided;
– with accidents which are expected to occur with an annual frequency above 10^{-3} to 10^{-4} and cause radiation exposures above the emergency level the existence of two independent effective barriers has to be proved.

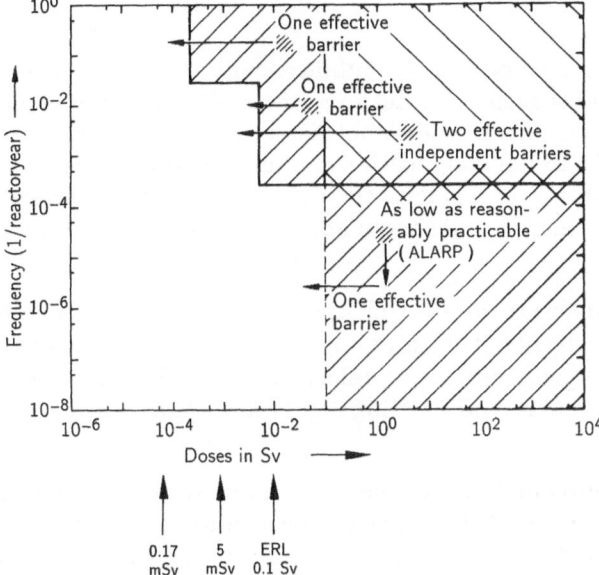

Fig. 6.6. Assessment reference levels and the effective barrier concept of the NII, United Kingdom

The barriers referred to are systems which either reduce the frequency of occurrence of an accident, mitigate its consequences or both. However, this concept of effective barriers is at present under revision.

To prove compliance with these requirements, reliability and safety analyses have to be presented in the licensing procedure in which the expected frequencies of occurrence and radiological consequences of accidents are assessed.

6.6.6 Health Risks of Individuals and Groups of Population

Safety goals concerning health risks are used, for example, in the Netherlands and the USA.

The criteria developed in the *Netherlands* [12] stipulate a limiting value for individual risk, i.e. an upper bound of 10^{-6} for the annual frequency of a person being killed who is permanently outdoors at a given distance from a source of hazard. Additionally, a minimum value of $10^{-8}\,a^{-1}$ is chosen. Between the two limiting values the ALARA principle will be applied. The same criteria are used for chemical hazard prevention (cf. [18]).

The safety goals in the USA comprise [13]:

- qualitative safety goals,
- quantitative safety goals.

The qualitative safety goals refer to individual and societal risks and contain the following requirements:

- individual risk: individual members of the public should be provided a level of protection from the consequences of nuclear power plant operation such that individuals bear no significant additional risk to life and health;
- societal risk: societal risks to life and health from nuclear power plant operation should be comparable to or less than the risks of generating electricity by viable competing technologies and should not be a significant addition to other societal risks.

The quantitative safety goals are targets which should be reached, if possible; they are not rigorous requirements. Thus a power station does not necessarily have to comply with all the goals. Given the uncertainties of risk assessment, the safety goals are not meant to replace existing regulations. The requirements, in detail, are as follows:

- the risk to an average individual in the vicinity of a nuclear power plant of prompt fatalities that might result from reactor accidents should not exceed one-tenth of one per cent (0.1 per cent) of the sum of prompt fatality risks resulting from other accidents to which members of the US population are generally exposed;
- the risk to the population in the area near a nuclear power plant of cancer fatalities that might result from nuclear power plant operation should not exceed one-tenth of one per cent (0.1 per cent) of the sum of cancer fatality risks resulting from all other causes.

The evaluation of the individual risks of prompt fatalities refers to an area within one mile of the nuclear power plant site boundary. For defining the population at risk a ten mile (16 km) radius is used.

6.6.7 Probabilistic Design Objectives on External Events and Fires

Probabilistic safety goals which refer to accident sequences, radiation exposure and risks also cover external events. However, specific safety goals for these events also exist in several countries, some of which were already in use before the safety objectives of the preceding sections were developed.

In this context design criteria which can be interpreted probabilistically should also be mentioned. These are:

- criteria for taking into account – or neglecting – rare events in the design of the plants on the basis of their frequencies of occurrence;
- criteria concerning load assumptions and design parameters on a probabilistic basis;
- criteria for taking into account secondary events based on their conditional probabilities of occurrence;
- criteria for taking into account or excluding combinations of events (e.g the simultaneous occurrence of external and plant internal events) based on their frequencies of occurrence.

It should be noted that probabilistic safety goals do not necessarily lead to identical requirements for design. Due to the site dependence of external events (type, intensity, frequency of occurrence) differing design requirements may result for comparable plants at different sites. In the following, design objectives for external events in some countries are reviewed.

By installing engineered safety features, *Belgian* [12] nuclear power stations are protected against aircraft crashes (civil and military), impacts from gas cloud explosions, the intrusion of toxic gases and large fires. Accidents resulting from external events are not considered if their expected annual frequency of occurrence is below 10^{-7}.

In *France* the general criterion for frequencies of occurrence of $< 10^{-7} \, \mathrm{a}^{-1}$ for groups of events with unacceptable consequences is also applied to external events and fires.

In particular, nuclear power stations are only designed to withstand crashes of small aircraft, since the following frequencies of aircraft crashes are calculated:

- small aircraft (< 5.7 tonnes) $\approx 10^{-6} \, \mathrm{a}^{-1}$ per site of a plant,
- big commercial aircraft $< 10^{-8} \, \mathrm{a}^{-1}$ per site of a plant,
- military aircraft $< 10^{-7} \, \mathrm{a}^{-1}$ per site of a plant.

In the design against other external events, for example impacts from industrial plants, the general criterion is applied. A possible endangering is analysed site-dependently with probabilistic methods. Probabilistic methods are also used for determining the seismic loads to be accounted for in the design of a nuclear power station.

In the *USA* probabilistic safety goals for seismic design are still under discussion, whereas those for aircraft crash, high tide and fires are already in use. In particular we have:

- Aircraft crash. Based on a realistic procedure it has to be demonstrated that the expected frequencies of unacceptable releases are less than $10^{-7} \, \mathrm{a}^{-1}$. For a more pessimistic assessment this value may be raised to $10^{-6} \, \mathrm{a}^{-1}$;

- High tide. Using a comprehensive site-dependent probabilistic approach, a 'maximum design high tide' is determined;
- Fires. The safety goals given in Table 6.7 apply.

In *Japan* no provisions against aircraft crash are made if the expected frequency of a crash is below $10^{-6}\,a^{-1}$.

In *Switzerland* and *Italy* the seismic design is largely based on probabilistic criteria which refer to site-specific load assumptions and design parameters.

Table 6.7. Probabilistic safety goals for fires

Possible consequences	Expected annual frequency
Fire without impact on safety relevant functions	$> 10^{-2}$
Fire with minimal impairment of safety relevant functions and a minimum loss of safety reserve	$10^{-2} \ldots 10^{-5}$
Fire causing the loss of the scram system and the residual heat removal	$10^{-5} \ldots 10^{-6}$
Fire causing abnormal release of radioactive substances	$< 10^{-6}$

6.6.8 Probabilistic Safety and Design Objectives in the Federal Republic of Germany

As already explained, the licensing procedure for nuclear installations in the FRG is basically deterministic and makes use of the concept of design basis accidents. In addition to rules and guidelines which specify safety factors and design criteria the 'Safety Criteria of the BMI' [3] stipulate that when examining the balance of the safety concept the reliability of safety relevant systems can be assessed probabilistically, if the state of science and technology permits this with sufficient accuracy. This complements the appraisal of the safety of nuclear power stations which uses deterministic methods.

A quantitative design goal was specified for the Thorium High Temperature Reactor (THTR). For a number of initiating events the product of their expected frequencies of occurrence and the unavailability of the residual heat removal systems was to be below 10^{-6} per year, if allowance is made for the emergency power and core-cooling systems and the long period of time during which the primary and secondary coolant circuits may remain uncooled.

Proposals which contain safety goals on radiation exposure can be found in a concept of the KTA [19]. Different event sequences are classified according to their frequencies of occurrence, and the corresponding admissible radiation exposures are specified. The following classification of event sequences is proposed.

- the event class 1 comprises normal operation and maintenance;

- event sequences are assigned to event class 2 if they are expected to occur during the lifetime of a plant. The frequency of such event sequences is assumed to be greater than $3 \cdot 10^{-2} \, a^{-1}$;
- the event class 3 contains event sequences which are not expected to occur in the lifetime of one plant but cannot be excluded during the lifetime of several plants. Their expected frequencies of occurrence range from $3 \cdot 10^{-2} \, a^{-1}$ to $1 \cdot 10^{-4} \, a^{-1}$;
- the event class 4 comprises the event sequences whose expected frequencies of occurrence are so low that they are not expected to occur in any plant. They are, however, used as limiting cases for the design of the safety systems of the plant. The frequency limits for such event sequences are between $10^{-4} \, a^{-1}$ and $10^{-5} \, a^{-1}$;
- the event class 5 contains those event sequences whose expected frequencies of occurrence are so low that they are not expected to occur in any of the plants and which have therefore not to be contemplated in the design according to §28 (3) of [2]. Nevertheless the plant must possess damage-mitigating properties in specific cases. The expected frequencies of occurrence of these event sequences are less than $10^{-5} \, a^{-1}$.

The protection goals are assigned to the event classes as follows:

- Event class 1: events belonging to this class allow operation to be continued without interruption. For this reason the limiting values of this class have to be fixed so that the loads remain confined to their operational values (including tolerances). In particular, the following is required:
  - The radiation exposure in the plant and its surroundings must remain within the admissible values for normal operation.
  - In order to ensure an incident-free and environmentally compatible operation the loads on components must remain within the limits authorized for normal operation; all process variables must remain within their operational ranges.

- Event class 2: if reactor operation is interrupted after an event sequence from this class, the immediate resumption of normal operation to the same extent as before the occurrence of the initiating event shall be possible after the removal of the cause of the disturbance. Therefore the limiting values of this class must be fixed in a way that the loads on components, with the exception of components subject to wear, are limited so that there is no doubt about their further utilization, even without additional tests and proofs. In particular, it is required that
  - the radiation exposure inside the plant and in its surroundings remains within the range permitted for normal operation,
  - loads which may threaten the integrity of barriers against radioactive releases remain within the range permitted for transients,
  - loads on the components of the safety relevant systems remain limited such that their integrity and function will not be impaired.

- Event class 3: event sequences of this class must be limited in such a way that the resumption of normal operation after the removal of the causes and consequences of the damage and after checking the state of the plant is possible. In particular, we have the following:

  - with respect to radiation exposure in the vicinity of the plant, §28 (3) of [2] is to be applied. Therefore the release of radioactive sustances has to be limited accordingly;
  - loads which can affect the integrity of the barriers against releases of radioactive substances must remain within the permitted range, which leaves a margin to the limiting values for failure;
  - for the important barriers which retain fission products this implies:
    1. The core geometry must be coolable.
    2. For light water reactors: the loads on the pressure boundary must remain within the range permitted for the operational condition C according to KTA 3201.2 [20] Sect. 3.3.3.4.
    3. For light water reactors: the loads on the containment are limited so that it can maintain its function as specified in KTA 3401.2 [21].

- Event class 4: after event sequences of this class resumption of normal operation is not possible in all cases. Therefore it is required that the safe shut-down of the plant, its power reduction to the long-term safe state and the long-term residual heat removal must be guaranteed. In particular:

  - radiation exposure of the surroundings must remain limited to the maximum values of §28 (3) of [2];
  - loads which can affect the integrity of the barriers against releases of radioactive substances must remain below the limiting values for failure;
  - for the important barriers which retain fission products this implies:
    1. The core geometry must be coolable.
    2. For light water reactors: the loads on the pressure boundaries must remain within the range permitted for the operational condition D according to KTA 3201.2 [20] Sect. 3.3.3.5.
    3. For light water reactors: the loads on the containment are limited so that it can maintain its function as specified in KTA 3401.2 [21].
  - A sufficient number of safety systems including their auxiliary systems must maintain their function to ensure long-term subcriticality, shut-down and long-term residual heat removal.

- Event class 5: due to their low probability of occurrence the plant is not required to cope with the consequences of event sequences of this class (§28 (3) of [2]). However, the plant must possess consequence-mitigating features in certain cases. This must be ensured by satisfying the specific requirements indicated for certain deterministically fixed event sequences.

Table 6.8 contains an assignment of the event classes to the definitions of the plant states.

Table 6.8. Assignment of the event classes to the definitions of the plant states (from [19])

	Assumed event sequence in the plant					
	Specified normal operation			'Incident'		Accident
	Maintenance	Normal operation	Operational disturbances			
Event class	1	2		3	4	5
Annual frequency			$> 3 \cdot 10^{-2}$	$3 \cdot 10^{-2}$ $\ldots 1 \cdot 10^{-4}$	$1 \cdot 10^{-4}$ $\ldots 1 \cdot 10^{-5}$	$< 1 \cdot 10^{-5}$

The safety requirements for external events in the FRG are essentially deterministic. Nevertheless events with very low frequencies of occurrence like aircraft crashes (approximately 10^{-5} to $10^{-7} \, a^{-1}$) are not included in the design basis accidents as specified in §28 (3) of [2]. Any measures against such events serve to reduce the population risk to a value below the general individual risks for health and life. They are implemented on the basis of judgment by the licensing authorities observing the principle of adequacy.

6.7 Outlook

About one hundred probabilistic studies of level 1 and above for plants in 23 countries have either been completed or are under way. There is a noticeable tendency towards unifying the boundary conditions and scope of the analyses, as reflected for example by the guidelines [22] and [23] and their application (cf. [24]). The guidelines are inscribed in programmes of periodic safety reviews which in some cases have already led to the so-called living PSAs (cf. [25]). These are probabilistic studies which serve to compare the effects of proposed alternatives for plant modifications and are updated regularly to reflect the status of the plant after implementation of the selected modification. In addition, they may be used to decide whether a plant should be shut down during the repair of failed systems by assessing the effect of the accompanying increase of unavailability on the risk of the entire plant.

A wider use of probabilistic analyses for decisions on proposed accident-management measures for the prevention of severe accidents or the mitigation of their consequences is foreseeable in the near future.

Methodology is also being improved, in particular as regards the treatment of human error and common-cause failures, the modelling of accident progression and the quantification of uncertainties.

All of this will assure an increased application of probabilistic safety analyses in support of licensing and optimization of the design and operation of nuclear power stations.

References

1. Gesetz über die friedliche Verwendung der Kernenergie und den Schutz gegen ihre Gefahren (Atomgesetz) vom 23. Dezember 1959 (BGBl. I, S. 814), in der Neufassung vom 15.7.1985 (BGBl I, 1565 ff), zuletzt geändert durch Art. 2 des Gesetzes vom 9.10.89 (BGBl. I, S. 1380) (English translation: GRS translations – safety codes and guides. Edition 2/87)
2. Verordnung über den Schutz vor Schäden durch ionisierende Strahlen (Strahlenschutzverordnung – StrlSchV) in der Bekanntmachung der Fassung vom 30. Juni 1989, Berichtigung vom 16. Oktober 1989 (English translation: GRS translations – safety codes and guides. Edition 3/89)
3. BMI-Sicherheitskriterien für Kernkraftwerke, vom 21. Oktober 1977, Bundesanzeiger 206 vom 3.11.1977 (English translation: GRS translations – safety codes and guides. Edition 13/78)
4. BMI Interpretationen zu den Sicherheitskriterien für Kernkraftwerke, Bekanntmachungen des BMI vom 17.5.1979, 28.11.1979, 2.3.1984 (English translation: GRS translations – safety codes and guides. Editions 9/80, 10/80, 2/84)
5. RSK-Leitlinien für Druckwasserreaktoren, 3. Ausgabe vom 14. Oktober 1981 (English translation: GRS translations – safety codes and guides. Edition 5/82)
6. Urteil des Verwaltungsgerichtshofs Baden-Württemberg, 30. März 1982 (Wyhl-Urteil) 575/77, 578/77, 583/77
7. Beschluss des Bundesverfassungsgerichts, 8.8.1978 (Kalkar Beschluss), 2 BvL 8/77
8. Der Bundesminister des Inneren: Leitlinien zur Beurteilung der Auslegung von Kernkraftwerken mit Druckwasserreaktor gegen Störfälle im Sinne des §28 Abs. 3 StrlSchV (Störfall-Leitlinien) v. 18. Oktober 1983. BAnz. Nr. 245 a v. 31.12.1983 (English translation: GRS translations – safety codes and guides. Edition 6/83)
9. Der Bundesminister des Inneren: Handbuch Reaktorsicherheit und Strahlenschutz. Loseblattsammlung, Sept. 1989
10. Merkpostenaufstellung mit Gliederung für einen Standardsicherheitsbericht für Kernkraftwerke mit Druckwasserreaktor oder Siedewasserreaktor, Bekanntmachung des BMI vom 26.7.1976 (GBMl. Nr. 26, 1976) (English translation: GRS translations – safety codes and guides. Edition 10/77)
11. Probabilistic safety criteria at the safety function/system level. IAEA-TEDOC-523, Vienna 1989
12. Summary report on safety objectives in nuclear power plants. EUR 12273EN. Brussels 1989
13. Nuclear Regulatory Commission: 10CFR part 50, Safety goals for the operation of nuclear power plants. Policy statement. August 1986
14. Design safety criteria for CEGB nuclear power stations. HS/R 167/81, London 1981
15. F.J. Remick: Letter to H.W. Zech, Jr. Chairman US NRC. 16.2.1989
16. Argentinian Atomic Energy Commission – CALIN 3.1.3 – criterios radiológicos relativos a accidentes, 1979
17. Argentinian Atomic Energy Commission – CALIN 3.2.2 – análisis de fallas para evaluación de riesgos, 1980
18. Kuijen, C.J. van: Management of industrial risks in the Netherlands – a probabilistic approach. The first international conference on safety and reliability objectives in high technology and industry. Amsterdam 2/3 March 1989
19. Statusbericht zum Konzept "Klassifizierung von Ereignisabläufen für die Auslegung von Kernkraftwerken" KTA-GS-47, 1985
20. KTA 3201.2 (Fassung 3.84): Komponenten des Primärkreises von Leichtwasserreaktoren, Teil 2: Auslegung, Konstruktion und Berechnung. BAnz. Nr.20a vom 30.1.1985

21. KTA 3401.2 (Fassung 6.85): Reaktorsicherheitsbehälter aus Stahl, Teil 2: Auslegung, Konstruktion und Berechnung. BAnz Nr. 203a vom 29.10.1985
22. Individual plant examination: submittal guidance. Final report. NUREG-1335. August 1989
23. Facharbeitskreis "Probabilistische Sicherheitsanalyse für Kernkraftwerke": Entwurf PSA-Leitfaden. (Draft of PSA guideline). Bonn November 1989
24. Polanski, X.: Implementation of the IPE methodology for five CECO plants. Proceedings PSA 89, International topical meeting – probability, reliability, and safety assessment. Pittsburgh 1989
25. Bonaca, M.V.: Utility use of PRA. Transactions of the 17th Water Reactor Safety Meeting. NUREG/CP-0104, October 1989

Appendix: Risk Values

Table A-1. Risk values for the generation of electricity from coal

Study Year of publication	[3] 1974				[4] 1976		[5] 1977	
Basis of comparison J	$2.36 \cdot 10^{16}$				$3.6 \cdot 10^{15}$		$2.21 \cdot 10^{16}$	
	Occupational accidents [a]		Harm to the public		Working time in 10^6 h	Occupational [b] deaths	Occupational accidents. Death	Harm to the public. Death
	Death	Injury	Death	Injury				
Mining	1.58	40.5	–	–	1.09	0.654	0.5	–
Preparation	–	–	–	–	–	–	–	–
Transport	0.055	5.1	0.55	1.17	0.33	0.112	–	1.3
Plant construction	–	–	–	–	0.2	0.014	0.05	–
Plant operation	0.03	1.2	–	–	0.2	0.005	–	–
Waste disposal					0.07	0.02	–	–
Sum	1.67	46.8	0.55	1.17	1.81	0.805	2 deaths (without accounting for flue gases) 2···100 deaths at 3% sulphur in coal	–
Sum normalized to $2.21 \cdot 10^{16}$ J	1.56	43.8	0.52	1.1	11.1	4.94	2 2···100	1.3

[a] Includes occupational diseases; in this study a death was valued with 6000 man-days lost.
[b] Includes occupational diseases.

Table A-1. (cont.)

Study	[6]
Year of publication	1978
Basis of comparison in J	$3.15 \cdot 10^{13}$

	Working time in h		Employees				Population			
			Accident		Disease		Accident		Disease	
	Construction	Operation	Death	Injury	Death	Disability	Death	Injury	Death	Disability
Mining Preparation }	49.6	1669	$0.7 \cdots 1.5 \cdot 10^{-3}$	$0.04 \cdots 0.07$	$0 \cdots 7.5 \cdot 10^{-4}$	$4.2 \cdots 8.4 \cdot 10^{-3}$	—	—	$1.4 \cdots 14 \cdot 10^{-3}$	—
Transport	26.5	92.1	$1.6 \cdots 5.0 \cdot 10^{-3}$	$1.3 \cdots 4.8 \cdot 10^{-2}$	—	—	$0.8 \cdots 1.9 \cdot 10^{-3}$	$1.6 \cdot 10^{-3}$	—	—
Plant construction Plant operation Waste disposal	429	407	$1.3 \cdots 9 \cdot 10^{-5}$	$1.6 \cdots 8.5 \cdot 10^{-3}$	—	—	—	—	$0.032 \cdots 0.095$	$190 \cdots 570$
Sum	505.1	2168.1	$2.3 \cdots 6.59 \cdot 10^{-3}$	$5.46 \cdots 12.65 \cdot 10^{-2}$	$0 \cdots 7.5 \cdot 10^{-4}$	$4.2 \cdots 8.4 \cdot 10^{-3}$	$0.8 \cdots 1.9 \cdot 10^{-3}$	$1.6 \cdot 10^{-3}$	$0.032 \cdots 0.109$	$190 \cdots 570$
Sum normalized to $2.21 \cdot 10^{16}$ J	$3.54 \cdot 10^{5}$	$1.52 \cdot 10^{6}$	$1.61 \cdots 4.62$	$33.3 \cdots 88.75$	$0 \cdots 0.53$	$2.95 \cdots 5.89$	$0.56 \cdots 1.33$	1.12	$23.4 \cdots 76.5$	$1.33 \cdots 3.99 \cdot 10^{5}$

Table A-1. (cont.)

Study / Year of publication	[7] 1980		[8] 1981		[9] 1981	[11] 1981			
Basis of comparison in J	$2.05 \cdot 10^{16}$		126 W		$3.6 \cdot 10^{15}$	$9.78 \cdot 10^{17}$			
	Occupational accidents		Occupational accidents		Occupational accidents.	Occupational Death accidents.	Permanent total disability	Permanent partial disability	Temporary total disability[d]
	Death[a]	Injury	Death	Injury[b]	Death[c]	Death			
Mining	$0.62 \cdots 1.0$	$42.5 \cdots 43$	$12 \cdot 10^{-9}$	$54 \cdot 10^{-6}$	–	–	–	–	–
Preparation	0.05	2.9	–	–	–	–	–	–	–
Transport	$0.3 \cdots 1.3$	$1.2 \cdots 5.9$	$9 \cdot 10^{-9}$	$43 \cdot 10^{-6}$	–	–	–	–	–
Plant construction	–	–	–	–	–	0.92	0	1.6	72.4
Plant operation	$0.1\ (0.02 \cdots 0.3)$	$3.3\ (2.7 \cdots 4.0)$	$10 \cdot 10^{-9}$	$67 \cdot 10^{-6}$	–	0	0	1	242
Waste disposal	–	–	–	–	–	–	–	–	–
Sum	$7.7 \cdots 9.1$	$50 \cdots 55$	$31 \cdot 10^{-9}$	$164 \cdot 10^{-6}$	0.99	0.92	0	2.6	314.4
Sum normalized to $2.21 \cdot 10^{16}$ J	$8.3 \cdots 9.8$	$54 \cdots 59$	Information for calculation missing		6.1	0.02	0	0.06	7.1

[a] An additional 6(0–30) deaths due to pollution are included (interval boundaries in brackets).

[b] In man-days lost; additionally, $430 \cdot 10^{-9}$ population deaths caused by air pollution; in addition persons killed by electricity, transport and desulphurization are indicated. Hours of operation are not given.

[c] In this reference death is considered equivalent to 6000 man-days lost.

[d] Values actually observed for Ontario Hydro which do not include disabilities of employees of suppliers. (Since only the category 'thermal' is indicated, it was assumed that coal is meant. Data on construction were converted to energy generation assuming a load factor of 0.7 and a useful plant life of 30 years.)

Table A-1. (cont.)

Study Year of publication	[12] 1981		[13]^b 1981			[15] 1981
Basis of comparison in J	$2.21 \cdot 10^{15}$		$2.04 \cdot 10^{16}$			$3.15 \cdot 10^{16}$
	Employees Death	Population Death	Employees Death	Occupational diseases	Population Death	Death by accident
Mining	–	–				1.4
Preparation	–	–				
Transport	0.08	0.01				0.2
Plant construction						
Plant operation	0.13 (3.6)	0.61 ⋯ 2.3 ^{a}				0.2
Waste disposal	–	–				
Sum	0.21 (4.4)	0.62 ⋯ 2.4	1.1	1 ⋯ 7	2.2 ⋯ 8.9	1.8
Sum normalized to $2.21 \cdot 10^{16}$ J	0.21 (4.4)	0.62 ⋯ 2.4	1.2	1.08 ⋯ 7.6	2.4 ⋯ 9.6	1.3

^{a} Values refer to the entire fuel cycle; numbers in brackets include events outside the economic region under consideration (Finland). Load factor is not indicated; therefore a value of 0.7 was assumed.
^{b} Death is considered equivalent to a loss of 6000 man-days and an occupational disease is assumed to imply a loss of 338 man-days.

Table A-2. Risk values for electricity generation from oil

Study Year of publication	[3] 1974	[4] 1976			[6] [a] 1978					
Basis of comparison in J	$2.36 \cdot 10^{16}$	$3.6 \cdot 10^{15}$			$3.15 \cdot 10^{13}$					
					Working time in h		Employees			
							Accident		Disease	
	Death	Injury	Working time in 10^6 h	Death	Con- struc- tion	Oper- ation	Death	Injury	Death	Disability
Extraction	0.063	7.5	0.09	0.004 ⎫	162	449	$0.14 \cdots 1.7 \cdot 10^{-3}$	$1.5 \cdots 12 \cdot 10^{-2}$	–	–
Processing	0.042	3.0	0.06	0.001 ⎬		52.3	$0.4 \cdots 1.4 \cdot 10^{-4}$	$1.6 \cdots 13 \cdot 10^{-3}$	–	–
Transport	0.03	1.1	0.15	0.052	3.39				–	–
Plant construction	–	–	0.16	0.01	–	–	–	–	–	–
Plant operation	0.037	1.5	0.11	0.004	250	226	$1.3 \cdots 5.0 \cdot 10^{-5}$	$0.9 \cdots 2 \cdot 10^{-3}$	–	–
Sum	0.172	13.1	0.57	0.071	415.4	727.3	$2.8 \cdots 20 \cdot 10^{-4}$	$4.7 \cdots 17 \cdot 10^{-3}$	–	–
Sum normalized to $2.21 \cdot 10^{16}$ J	0.16	12.3	3.5	0.44	$2.9 \cdot 10^5$	$5.1 \cdot 10^5$	$0.2 \cdots 1.4$	$3.3 \cdots 11.9$	–	–

[a] The sums were taken over directly from reference [6]; they do not necessarily result from adding up the individual contributions.

Table A-2. (cont.)

Study	[6]	[7]	[9]	[13]	[15]
Year of publication	1978	1980	1981	1981	1981
Basis of comparison in J	$3.15 \cdot 10^{13}$	$1.06 \cdot 10^{18}$	$3.6 \cdot 10^{15}$	$2.04 \cdot 10^{16}$	$3.15 \cdot 10^{16}$

	[6] Population Accident Death	[6] Population Accident Injury	[6] Population Disease Death	[6] Population Disease Disability	[7] Death	[7] Disability	[9] Death	[13] Employees Death	[13] Population Death	[15] Population Death	[15] Population Death
Extraction }	–	–	–	–	–	–	–	–	–	0.3	–
Processing }	–	–	–	–	–	–	–	–	–	not available	–
Transport	–	–	–	–	–	–	–	–	–	–	–
Plant construction	–	–	–	–	–	–	–	–	–	–	–
Plant operation	–	–	$0.012\cdots0.035$	$70\cdots210$	–	–	–	–	–	not registered	–
Sum	–	–	$0.012\cdots0.035$	$70\cdots210$	$88\cdots4400$	$4000\cdots79\,000$	$0.072\,^a$	0.024	0.02	$3.9\cdots15.6$	0.3
Sum normalized to $2.21 \cdot 10^{16}$ J	–	–	$8.4\cdots24.5$	$4.9\cdots14.7\cdot10^{4}$	$1.8\cdots92$	$83\cdots1647$	0.44	0.026	0.022	$4.2\cdots16.9$	0.21

^{a} In this study death was considered equivalent to 6000 days of work lost.

Table A-3. Risk values for electricity generation from natural gas

Study[a] Year of publication	[3] 1974	[6][b] 1978	[7] 1980		
Basis of comparison in J	$2.36\cdot10^{16}$	$3.15\cdot10^{13}$	$1.07\cdot10^{18}$		

	Occupational accidents		Working time in h		Employees		Population		Disease due to emissions from plant construction	Death	Disability
					Accident		Accident				
	Death	Injury	Construction	Operation	Death	Injury	Death	Injury			
Extraction	0.021	2.5	49	449	$3.9\cdots31\cdot10^{-5}$	$0.004\cdots0.031$	—	—			
Processing	0.006	0.56	included in 'extraction'								
Transport	0.024	1.3	3.39	52.3	$2.9\cdots3.4\cdot10^{-5}$	$0.0017\cdots0.0019$	—	—			
Plant construction			250						Death $0.5\cdots1.4\cdot10^{-5}$		
Plant operation	0.037	1.5		226	$1.4\cdots5.3\cdot10^{-5}$	$0.9\cdots2.1\cdot10^{-3}$	$9\cdot10^{-6}$	$4.7\cdot10^{-5}$	Invalidity $0.028\cdots0.85$		
Waste disposal											
Sum	0.088	5.86	302.4	727.3	$16\cdots48\cdot10^{-5}$	$0.009\cdots0.036$	$9\cdot10^{-6}$	$4.7\cdot10^{-5}$	$0.5\cdots1.4\cdot10^{-5}$ $0.028\cdots0.85$	6	600
Sum normalized to $2.21\cdot10^{16}$ J	0.082	5.48	$2.12\cdot10^{5}$	$5.1\cdot10^{5}$	$0.11\cdots0.34$	$6.3\cdots25.3$	$6.3\cdot10^{-3}$	$3.3\cdot10^{-2}$	$3.5\cdot10^{-3}\cdots9.8\cdot10^{-3}$ $19.6\cdots596$	0.12	12.4

[a] The values for natural gas in reference [4] are identical with those for oil (cf. Table A-2).
[b] The sums were taken over directly from reference [6]; they do not necessarily result from adding up the individual contributions.

Table A-4. Risk values for electricity generation from nuclear fission in LWRs

Study Year of publication	[3] 1974				[3] 1974 (BWR)				[4] 1976		[5] 1977	
Basis of comparison in J	$2.36 \cdot 10^{16}$				$2.36 \cdot 10^{16}$ (BWR)				$3.6 \cdot 10^{15}$		$2.21 \cdot 10^{16}$	
	Occupational accidents[a]		Harm to the public		Occupational accidents[a]		Harm to the public		Working time in 10^6 h	Death[b]	Occupational accidents. Death[c]	Harm to the public. Death
	Death	Injury	Death	Injury	Death	Injury	Death	Injury				
Mining	0.09	3.6	–	–	0.09	3.3	–	–	0.03	0.028	0.26	0.08
Preparation	0.005	1.4	–	–	0.005	1.4	–	–	0.002	0.0002	0.031	0.02
Transport	0.002	0.12	0.009[a]	0.08[a]	0.002	0.14	0.009[a]	0.08[a]	0.0008	0.0001	0.02	0.01
Plant construction	0.01	1.3	–	–	0.01	1.3	–	–	0.24	0.017	0.07···0.2	–
Plant operation	–	–	–	–	–	–	–	–	0.035	0.0014	0.15	0.033
Reprocessing	included in 'transport'		–	–	included in 'transport'		–	–	0.004	0.0002	0.004	0.08
Waste disposal	–	–	–	–	–	–	–	–	0.001	0.0003	–	–
Sum	0.11	6.42	0.009	0.08	0.11	6.14	0.009	0.08	0.31	0.047	0.54···0.67	0.22
Sum normalized to $2.21 \cdot 10^{16}$ J	0.1	6.0	0.008	0.07	0.1	5.7	0.008	0.07	1.9	0.29	0.54···0.67	0.22

[a] Excluding effects of radiation exposure.
[b] Including occupational diseases.
[c] Including late fatalities caused by radiation.

Table A-4. (cont.)

| Study Year of publication | | [6][b] 1978 | | | | | | | | |
| Basis of comparison in J | | $3.15 \cdot 10^{13}$ | | | | | | | | |

| | Working time in h | | Employees | | | | Population | | | |
| | | | Accident | | Disease | | Accident | | Disease | |
	Con-struction	Oper-ation	Death	Injury	Death	Disability	Death	Injury	Death	Disability
Mining Preparation }	21.3	205	$0.8 \cdots 5.7 \cdot 10^{-4}$	$3.4 \cdots 16 \cdot 10^{-3}$	$2.2 \cdots 60 \cdot 10^{-5}$	$1.1 \cdots 1.6 \cdot 10^{-5}$	–	– Emissions (SO_x)	$3.1 \cdot 10^{-5}$ Emissions $7 \cdots 20 \cdot 10^{-8}$	– / $0.43 \cdots 0.128$
Transport Plant construction	– / –	– / –	$2.7 \cdots 12 \cdot 10^{-6}$	$6 \cdots 20 \cdot 10^{-5}$	$0.3 \cdots 40 \cdot 10^{-7}$	$0.6 \cdots 80 \cdot 10^{-7}$	$1.2 \cdot 10^{-5}$	$1.1 \cdot 10^{-4}$	$22 \cdot 10^{-4}$	$44 \cdot 10^{-8}$
Plant operation	604	250	$1.3 \cdots 1.7 \cdot 10^{-5}$	$1.7 \cdot 10^{-3}$	$7.4 \cdot 10^{-5}$	$6 \cdots 22 \cdot 10^{-5}$	–	–	$3 \cdots 23 \cdot 10^{-5}$	$7 \cdots 260 \cdot 10^{-7}$
Reprocessing Waste disposal } }	8	31	$2.0 \cdot 10^{-7}$	$2.7 \cdot 10^{-4}$	$7.4 \cdot 10^{-5}$	$1.5 \cdot 10^{-4}$	–	–	$2.5 \cdot 10^{-4}$	$1.5 \cdot 10^{-4}$
Sum	633.3	486	$2.5 \cdots 7.5 \cdot 10^{-4}$	$16 \cdots 29 \cdot 10^{-3}$	$1.3 \cdots 7.2 \cdot 10^{-4}$	$2.2 \cdots 3.9 \cdot 10^{-4}$	$1.2 \cdot 10^{-5}$	$1.1 \cdot 10^{-4}$	$3.1 \cdots 5.1 \cdot 10^{-4}$	$0.043 \cdots 0.128$
Sum normalized to $2.21 \cdot 10^{16}$ J	$4.4 \cdot 10^5$	$3.4 \cdot 10^5$	$0.18 \cdots 0.53$	$11.2 \cdots 20.3$	$9.1 \cdots 50.5 \cdot 10^{-2}$	$1.5 \cdots 2.7 \cdot 10^{-1}$	$8.4 \cdot 10^{-3}$	$7.7 \cdot 10^{-2}$	$0.22 \cdots 0.36$	$30.2 \cdots 89.8$

[a] Including late fatalities caused by radiation.
[b] The sums were taken over directly from reference [6]; they do not necessarily result from adding up the individual contributions.

Table A-4. (cont.)

Study Year of publication	[7] 1980				[9]ᵃ 1981	[11] 1981			
Basis of comparison in J	$2.05 \cdot 10^{16}$				$3.6 \cdot 10^{15}$	$5.58 \cdot 10^{17}$			
	Employees		Population		Employees Death	Employees Death	Permanent total disability	Permanent partial disability	Temporary total disability ᵇ
	Death	Injury	Death	Disability					
Mining	0.44	$12.33 \cdots 14.79$	0.08	0.08	–	–	–	–	–
Preparation	0.038	1.334	0.002	0.002	–	–	–	–	–
Transport	$9.39 \cdot 10^{-3}$	$9.93 \cdot 10^{-2}$	$6.1 \cdot 10^{-4}$	$6.1 \cdot 10^{-4}$	–	–	–	–	–
Plant construction	–	–	–	–	–	0.72	0	0.24	118
Plant operation	0.183	1.2	0.017	0.017	–	0	0	3	113
Reprocessing Waste disposal }	$5.75 \cdot 10^{-3}$	$7.07 \cdot 10^{-2}$	$5.1 \cdot 10^{-5}$	$5.1 \cdot 10^{-5}$	–	–	–	–	–
Sum	0.68	$15 \cdots 17.5$	0.1	0.1	0.23	0.72	0	3.24	231
Sum normalized to $2.21 \cdot 10^{16}$ J	0.73	$16.2 \cdots 18.9$	0.11	0.11	1.4	0.03	0	0.13	9.15

ᵃ In this study death was considered equivalent to 6000 man-days lost.
ᵇ Values actually observed for Ontario Hydro which do not include disabilities of employees of suppliers. Data on construction were converted to energy generation assuming a load factor of 0.7 and a useful plant life of 30 years.

Table A-4. (cont.)

Study Year of publication	[12]ᵃ 1981		[13]ᵇ 1981		[15] 1981	[17] 1981	
Basis of comparison in J	2.21·10¹⁶		2.04·10¹⁶ PWR		3.15·10¹⁶	3.15·10¹⁶	
	Employees Death	Population Death	Employees Death	Employees Disease	Employees Death	Employees Death	Harm to the public. Death
Mining	–	–	–	–	0.1	–	–
Preparation	0.001	–	–	–	–	–	–
Transport	–	–	–	–	negligible	–	–
Plant construction			–	–	–	–	–
Plant operation	0.16 (0.4)	0.024 (0.5)	–	–	} 0.15	–	–
Reprocessing	}	0.006 (0.1)	–	–		–	–
Waste disposal	–		–	–	–	–	–
Sum	0.161 (0.401)	0.03 (0.6)	0.246	0.46	0.25	0.24···1.2	0.03···0.18
Sum normalized to 2.21·10¹⁶ J	0.161 (0.401)	0.03 (0.6)	0.266	0.498	0.18	0.17···0.84	0.021···0.13

ᵃ A load factor of 0.7 was assumed; values in brackets refer to events outside the economic region under consideration (Finland). The values were in part assigned to power plant operation, since no detailed assignments are given; however, they are believed to contain contributions from other steps of the fuel cycle as well.
ᵇ Includes theoretical radiation damage; 0.0031 population fatalities have to be added to the values indicated.

Table A-5. Risk values for electricity generation from solar heat

		Employees				Population				[6]^a 1978 / [9] 1981
Study / Year of publication										[6]^a 1978 [9] 1981
Basis of comparison										3.15·10^{13} 3.6·10^{15}
	Working time in h	Accident		Disease		Accident		Disease		Death
		Death	Injury	Death	Disability	Death	Injury	Death	Disability	
Transport	–	1.4···4.5·10^{-4}	1.2···4.3·10^{-3}	–	–	0.7···1.7·10^{-4}	1.4·10^{-4}	–	–	–
Plant construction	8730	2.7·10^{-3}	0.1	included in 'accident'	0.0028	–	–	–	–	–
Emissions from steel and aluminium production	–	–	–	–	–	–	–	2.1···6.2·10^{-4}	1.2···3.7	–
Plant operation	–	0.83·10^{-3}	0.025	4.4·10^{-6}	1·10^{-4}	–	–	–	–	–
Sum	8730	3.67···3.98·10^{-3}	0.126···0.129	4.4·10^{-6}	2.9·10^{-3}	0.7···1.7·10^{-4}	1.4·10^{-4}	2.1···6.2·10^{-4}	1.2···3.7	0.43^b
Sum normalized to 2.21·10^{16} J	6.12·10^6	2.6···2.8	88.5···90.7	3.09·10^{-3}	2.0	4.9···11.9·10^{-2}	9.8·10^{-2}	0.15···0.43	842···2595	2.64

^{a} Without support and storage systems.
^{b} 37% of this value correspond to the construction of the plant.

Table A-6. Risk values for electricity generation from photovoltaic conversion

Study [a] Year 1978 of publication	Work in man-hours	Employees				Population			
Basis of comparison in J $3.15 \cdot 10^{13}$		Accident		Disability	Disease	Accident		Disease	
		Death	Injury		Death	Death	Injury	Death	Disability
Transport	–	$3.2 \cdots 10 \cdot 10^{-4}$	$2.6 \cdots 9.6 \cdot 10^{-3}$	–	included in 'accident'	$1.6 \cdots 3.8 \cdot 10^{-4}$	$3.2 \cdot 10^{-4}$	–	–
Plant construction	$10\,810 \cdots 20\,810$	$5.9 \cdots 9.0 \cdot 10^{-4}$	$0.05 \cdots 0.08$	$0.002 \cdots 0.004$	–	–	–	–	–
Emissions from steel and aluminium production	–	–	–	–	–	–	–	$1.5 \cdots 4.5 \cdot 10^{-3}$	$9.0 \cdots 27$
Plant operation	–	$0.83 \cdot 10^{-3}$	0.025	$1.0 \cdot 10^{-4}$	$4 \cdot 10^{-6}$	–	–	–	–
Sum	$10\,810 \cdots 20\,810$	$1.74 \cdots 2.73 \cdot 10^{-3}$	$7.76 \cdots 11.5 \cdot 10^{-2}$	$2.1 \cdots 4.1 \cdot 10^{-3}$	$4 \cdot 10^{-6}$	$1.6 \cdots 3.8 \cdot 10^{-4}$	$3.2 \cdot 10^{-4}$	$1.5 \cdots 4.5 \cdot 10^{-3}$	$9.0 \cdots 27$
Sum normalized $7.6 \cdot 10^{6}$ to $2.21 \cdot 10^{16}$ J	$7.6 \cdot 10^{6}$ $\cdots 1.5 \cdot 10^{7}$	$1.22 \cdots 1.91$	$54.4 \cdots 80.7$	$1.47 \cdots 2.88$	$2.8 \cdot 10^{-3}$	$0.11 \cdots 0.27$	0.22	$1.05 \cdots 3.16$	6314 $\cdots 1.89 \cdot 10^{4}$

[a] Without the energy storage and support systems treated in the study.

Table A-6. (cont.)

Study Year of publication	[9]a 1981	[16] 1981		[17] 1981	
Basis of comparison in J	$3.6 \cdot 10^{15}$	$1.06 \cdot 10^{15}$		$3.15 \cdot 10^{16}$	
	Employees Death	Working time b in 10^6 h	Employees Death	Employees Death	Population Death
Transport	—	—	—	—	—
Plant construction	—	—	—	—	—
Emissions from steel and aluminium production	—	—	—	—	—
Plant operation	—	—	—	—	—
Sum	0.25	0.54	0.023	0.43···0.73	U
Sum normalized to $2.21 \cdot 10^{16}$ J	1.53	11.26	0.48	0.30···0.51	U

a In this study death is considered equivalent to 6000 working days lost; 37% of the indicated values refer to the construction of the plant.
b Using 2000 h per man-year.
U = unknown or negligible.

Table A-7. Risk values for solar domestic heating and warm water preparation

		Employees				Population				
Study		[6][c] 1978								[16][a] 1981
Year of publication										
Basis of comparison in J thermal energy		$3.15 \cdot 10^{13}$								$1.06 \cdot 10^{15}$
	Working time in h	Accident		Disease		Accident		Disease		Working time in 10^6 h[b] / Death
		Death	Injury	Death	Injury	Death	Injury	Death	Disability	
Transport	–	$3.2 \cdots 10 \cdot 10^{-4}$	$0.003 \cdots 0.011$	–	–	$1.6 \cdots 4.0 \cdot 10^{-4}$	$3.3 \cdot 10^{-4}$	–	–	–
Plant construction	29 642	0.003	0.15	included in 'accident'	0.044	–	–	–	–	–
Emissions from steel and aluminium production	–	–	–	–	–	–	–	$3.2 \cdots 9.7 \cdot 10^{-4}$	$1.9 \cdots 5.9$	–
Plant operation	–	$2.1 \cdots 4.7 \cdot 10^{-3}$	$0.08 \cdots 0.15$	$1.3 \cdots 2.9 \cdot 10^{-5}$	$3.0 \cdots 6.8 \cdot 10^{-5}$	–	–	–	–	–
Sum	29 642	$5.1 \cdots 8.5 \cdot 10^{-3}$	$0.23 \cdots 0.33$	$1.3 \cdots 3.8 \cdot 10^{-5}$	0.044	$2.0 \cdots 4.6 \cdot 10^{-4}$	$3.8 \cdots 6.8 \cdot 10^{-4}$	$7.4 \cdots 23 \cdot 10^{-4}$	$4.1 \cdots 12$	0.59 0.0425

[a] Arithmetic mean of the values for active solar domestic heating and warm water preparation.

[b] Conversion: 1 man-year $\hat{=}$ 2000 h.

[c] The sums were taken over directly from reference [6]; they do not necessarily result from adding up the individual contributions.

Table A-8. Risk values electricity generation from wind energy

Study	[6]^a								
Year of publication	1978								
Basis of comparison in J	3.15 · 10^13								

	Working time^b in h	Employees				Population			
		Accident		Disease		Accident		Disease	
		Death	Injury	Death	Disability	Death	Injury	Death	Disability
Transport	—	$2.0 \cdots 6.3 \cdot 10^{-4}$	$1.5 \cdots 6.0 \cdot 10^{-3}$	included in 'accident'	—	$1.0 \cdots 2.4 \cdot 10^{-4}$	$2.0 \cdot 10^{-4}$	—	—
Plant construction	5159 ⋯ 7659	$3.7 \cdots 7.4 \cdot 10^{-4}$	$0.03 \cdots 0.04$		$7 \cdots 9 \cdot 10^{-4}$	—	—	—	—
Emissions from steel and aluminium production	—	—	—	—	—	—	—	$1.2 \cdots 3.6 \cdot 10^{-4}$	$0.73 \cdots 2.2$
Plant operation	—	$2.6 \cdot 10^{-3}$	0.079	$1.4 \cdot 10^{-5}$	$3.2 \cdot 10^{-4}$	—	—	—	—
Sum	5159 ⋯ 7659	$3.2 \cdots 4.0 \cdot 10^{-3}$	$0.11 \cdots 0.13$	$1.4 \cdot 10^{-5}$	$1.02 \cdots 1.22 \cdot 10^{-3}$	$1.0 \cdots 2.4 \cdot 10^{-4}$	$2.0 \cdot 10^{-4}$	$1.2 - 3.6 \cdot 10^{-4}$	$0.73 \cdots 2.2$
Sum normalized to $2.21 \cdot 10^{16}$ J	$3.6 \cdots 5.4 \cdot 10^{6}$	$2.2 \cdots 2.8$	77 ⋯ 87.7	$9.8 \cdot 10^{-3}$	$0.71 \cdots 0.86$	$0.07 \cdots 0.17$	0.14	$0.084 \cdots 0.25$	512 ⋯ 1543

^a Without support and storage systems.
^b Conversion with 2000 h per man-year.

Table A-8. (cont.)

Study Year of publication	[7] 1980			[8]^b 1981		[16] 1981	
Basis of comparison in J	$1.06 \cdot 10^{15}$			3.9 W		$1.06 \cdot 10^{15}$	
	Working^a time in h	Death	Disease	Death	Injury in man-days lost	Working time^c in h	Death
Transport	–	–	–	–	–	–	–
Plant construction	$6.2 \cdot 10^5$	$2 \cdot 10^{-2}$	24.3	–	–	–	–
Emissions from steel and aluminium production	–	–	–	$3 \cdot 10^{-9}$	$29 \cdot 10^{-6}$	–	–
Plant operation	$1.1 \cdot 10^5$	$2.7 \cdot 10^{-3}$	3.3	$3 \cdot 10^{-9}$	$29 \cdot 10^{-6}$	–	–
Sum	$7.3 \cdot 10^5$	$2.27 \cdot 10^{-2}$	27.6	–	–	$4.2 \cdot 10^5$	0.018
Sum normalized to $2.21 \cdot 10^{16}$ J	$1.5 \cdot 10^7$	0.47	575.4	–	–	$8.8 \cdot 10^6$	0.38

^{a} Conversion with 2000 h per man-year.
^{b} Additionally $4 \cdot 10^{-9}$ deaths and $20 \cdot 10^{-6}$ man-days lost for energy transport. The conversion base injury – man-days lost is not indicated.
^{c} Assumption: 1 man-year $\cong$ 2000 h.

Table A-9. Risk values for electricity generation from oceanic heat

Study	[6]ᵃ
Year of publication	1978
Basis of comparison in J	$3.15 \cdot 10^{13}$

	Working time in h	Employees				Population			
		Accident		Disease		Accident	Disease		
		Death	Injury	Death	Disability	Death	Injury	Death	Disability
Transport	—	$0.6 \cdots 1.9 \cdot 10^{-4}$	$5.1 \cdots 19 \cdot 10^{-4}$	—	—	$3.1 - 7.4 \cdot 10^{-5}$	$0.6 \cdot 10^{-4}$	—	—
Plant construction	$8316 \cdots 10716$	$9.4 \cdots 13 \cdot 10^{-3}$	$0.06 \cdots 0.08$	included in 'accident'	—	—	—	—	—
Emissions from steel and aluminium production	—	—	—	—	—	—	—	$8 \cdots 25 \cdot 10^{-5}$	$0.5 \cdots 1.5$
Plant operation	—	$4.25 \cdot 10^{-4}$	0.014	$2.7 \cdot 10^{-6}$	$5.7 \cdot 10^{-5}$	—	—	—	—
Sum	$8316 \cdots 10716$	$1.4 \cdots 1.9 \cdot 10^{-3}$	$0.07 \cdots 0.09$	$2.7 \cdot 10^{-6}$	$2.7 \cdots 3.5 \cdot 10^{-3}$	$3.1 \cdots 7.4 \cdot 10^{-5}$	$0.6 \cdot 10^{-4}$	$8 \cdots 25 \cdot 10^{-5}$	$0.5 \cdots 1.5$
Sum normalized to $2.21 \cdot 10^{16}$ J	$5.8 \cdots 7.5 \cdot 10^{6}$	$0.98 \cdots 1.33$	$49.1 \cdots 63.1$	$1.9 \cdot 10^{-3}$	$1.89 \cdots 2.46$	$2.17 \cdots 5.19 \cdot 10^{-2}$	$4.2 \cdot 10^{-2}$	$5.6 \cdots 17.5 \cdot 10^{-2}$	$351 \cdots 1052$

ᵃ The sums were taken over directly from reference [6]; they do not necessarily result from adding up the individual contributions.

Table A-10. Risk value for electricity generation from methanol

Study	[6]								
Year of publication	1978								
Basis of comparison in J	$3.15 \cdot 10^{13}$								

	Working time in h	Employees				Population			
		Accident		Disease		Accident		Disease	
		Death	Injury	Death	Disability	Death	Injury	Death	Disability
Transport	–	$3.3 \cdot 10^{-4}$	$2.8 \cdots 10 \cdot 10^{-3}$	–	–	$1.7 \cdots 4 \cdot 10^{-4}$	$3.3 \cdot 10^{-4}$	–	–
Plant construction	–	$1.6 \cdots 2.5 \cdot 10^{-3}$	$0.1 \cdots 0.12$	included in 'accident'	0.012	–	–	–	–
Emissions from the acquisition of building materials and fuel combustion	–	–	–	–	–	–	–	$2.9 \cdots 18 \cdot 10^{-4}$	$1.7 \cdots 11$
Plant operation	–	$1.5 \cdot 10^{-2}$	0.55	$9 \cdot 10^{-5}$	$2.1 \cdot 10^{-3}$	–	–	–	–
Sum	–	$1.7 \cdots 1.8 \cdot 10^{-2}$	$0.65 \cdots 0.68$	$9 \cdot 10^{-5}$	$1.4 \cdot 10^{-2}$	$1.7 \cdots 4 \cdot 10^{-4}$	$3.3 \cdot 10^{-4}$	$2.9 \cdots 18 \cdot 10^{-4}$	$1.7 \cdots 11$
Sum normalized to $2.21 \cdot 10^{16}$ J	–	$11.9 \cdots 12.5$	$456 \cdots 477$	$6.3 \cdot 10^{-2}$	9.8	$0.12 \cdots 0.28$	0.23	$0.2 \cdots 1.26$	$1193 \cdots 7717$

Table A-11. Risk values for electricity generation from hydraulic energy

Study	[4]	[6]
Year of publication	1976	1978
Basis of comparison in J	$3.6 \cdot 10^{15}$	$3.15 \cdot 10^{13}$

	Working time in h	Death during work^a	Work in man-hours	Employes				Population			
				Accident		Disease		Accident		Disease	
				Death	Injury	Death	Disability	Death	Injury	Death	Disability
Plant construction	0.1	0.068	511.4^b	$8.9 \cdots 20 \cdot 10^{-4}$	0.34	—	$8 \cdot 10^{-5}$	—	—	—	—
Dam rupture	—	—	—	—	—	—	—	$1.1 \cdots 1.6 \cdot 10^{-3}$	$1.1 \cdots 12.8 \cdot 10^{-3}$	—	—
Plant operation	0.11	0.004	—	$6.3 \cdot 10^{-4}$	0.013	—	—	—	—	—	—
Sum	0.21	0.012	511.4	$1.5 \cdots 2.6 \cdot 10^{-3}$	0.35	—	$8 \cdot 10^{-5}$	$1.1 \cdots 1.6 \cdot 10^{-3}$	$1.1 \cdots 12.8 \cdot 10^{-3}$	$2.2 \cdots 6.6 \cdot 10^{-6}$	$0.013 \cdots 0.039$
Sum normalized to $2.21 \cdot 10^{16}$ J	1.29	0.074	$3.6 \cdot 10^5$	$1.05 \cdots 1.82$	246	—	$5.6 \cdot 10^{-2}$	$0.77 \cdots 1.12$	$0.77 \cdots 8.98$	$1.54 \cdots 4.63 \cdot 10^{-3}$	$9.1 \cdots 27.4$

^{a} Includes consequences of occupational diseases.
^{b} Only material acquisition without power plant construction.

Table A-11. (cont.)

Study Year of publication	[11] [a] 1981			
Basis of comparison in J	$1.3 \cdot 10^{18}$			
	Death	Permanent total disability	Permanent partial disability	Temporary total disability
Plant construction	–	–	–	–
Dam rupture	–	–	–	–
Plant operation	1	not occurred	1	38
Sum	1	0	1	38
Sum normalized to $2.21 \cdot 10^{16}$ J	$1.7 \cdot 10^{-2}$	0	$1.7 \cdot 10^{-2}$	0.65

[a] Values actually observed for Ontario Hydro.

Table A-12. Risk values for several non-conventional systems for generating electricity (Average number of fatalities per $3.5 \cdot 10^{16}$ J of generated energy (from [17]) 1981)

System of energy conversion	Fast breeder reactor	Solar energy from satellites	Decentralized photovoltaic energy	Coal gasification with gas turbine	Nuclear fusion
Group of risk					
Population	0.03···0.18	U	U	5.4···76	0.0001
Employees	0.21···0.94	0.26···0.67	1.92···4.39	1.3···3.1	0.22···0.44
Activity					
Material acquisition and construction[a]	0.12···0.20	0.19···0.55	1.04···1.94	0.11···0.18	0.16···0.38
Operation and maintenance	0.12···0.92	0.07···0.12	0.88···2.45	6.5···79	0.03···0.06
Cause					
Accidents and diseases not produced by radiation	0.17···0.51	0.26···0.67	1.9···4.4	6.6···79	0.22···0.44
Radiation	0.07···0.61	U	U	0.0023	U
Sum	0.24···1.1	0.26···0.67	1.92···4.4	6.6···79	0.22···0.44
Sum normalized to $2.21 \cdot 10^{16}$	0.15···0.69	0.16···0.42	1.21···2.78	4.17···49.9	0.14···0.28

[a] Averaged over a lifetime of 30 years.
U = unknown or negligible.

Table A-13. Risk values for different systems using renewable energy sources. (Volume of work and risk data for the use of different systems of energy generation based on $1.06 \cdot 10^{15}$ J (from [16]) 1981)

System of energy conversion	Passive solar heating	Active solar heating and cooling	Solar industrial process heat	Anaerobic fermentation	Wood pyrolisis	Decentralized wind energy without storage	Decentralized wind energy with storage
Working time in 10^6 h	0.28	1.04	0.56	0.17	0.11	2	2.6
Deaths	0.015	0.057	0.033	0.017	0.0048	0.087	0.11
Accidents and diseases in man-days lost	110	440	250	78	78	1000	1200
Fatalities normalized to $2.21 \cdot 10^{16}$ J	0.31	1.19	0.69	0.35	0.10	1.81	2.29
Working days lost due to accidents and disease normalized to $2.21 \cdot 10^{16}$ J	2293	9173	5212	1626	1626	20 849	25 018

Table A-14. Risk values for decentralized photovoltaic energy supply

Study Year of publication	[16] 1981		[17] 1981
Basis of comparison in J	$1.06 \cdot 10^{15}$		$3.5 \cdot 10^{16}$
	Working time^a in 10^6 h	Employees		Death
		Death	Disease^b	
Sum	3	0.17	1200	$1.92 \cdots 4.4$
Sum normalized to $2.21 \cdot 10^{16}$ J	62.5	3.5	$2.5 \cdot 10^4$	$1.21 \cdots 2.78$

^{a} Conversion: 1 man-year $\widehat{=}$ 2000 h.
^{b} In working days lost.

Table A-15. Risk values for transmission lines [a]

Study	[6] [b]								
Year of publication	1978								
Basis of comparison	1000 km per year over system lifetime (appr. 30 years)								

	Working hours	Employees				Population			
		Accident		Disease		Accident		Disease	
		Death	Injury	Death	Disability	Death	Injury	Death	Disability
Transportation	–	0.005···0.015 (0.012···0.038) (0.013···0.042)	0.033···0.13 (0.10···0.37) (0.11···0.40)	–	–	0.0022 (0.005) (0.006)	0.005 (0.011) (0.012)	–	–
Construction	$1.19 \cdot 10^7$ $(2.67 \cdot 10^7)$	0.014 (0.039) (0.043)	1.33 (3.70) (4.10)	included in 'accident'	0.067 (0.184) (0.204)	–	–	–	–
Emissions of steel production	–	–	–	–	–	–	–	0.0034···0.01 (0.011···0.033) (0.012···0.037)	20.1···60 (67···200) (73···220)
Operation	–	0.0021	0.065	–	–	–	–	–	–
Sum	$1.19 \cdot 10^7$ $(2.67 \cdot 10^7)$	0.021···0.031 (0.053···0.079) (0.0058···0.087)	1.43···1.54 (3.9···4.1) (4.3···4.6)	–	0.069 (0.186) (0.206)	0.0022 (0.005) (0.006)	0.005 (0.011) (0.012)	0.0034···0.01 (0.011···0.033) (0.012···0.037)	20.1···60 (67···200) (73···220)

230 kV
[a] Lines for (500) kV
(765) kV
[b] The sums were taken over directly from reference [6]; they do not necessarily result from adding up the individual contributions.

Subject Index